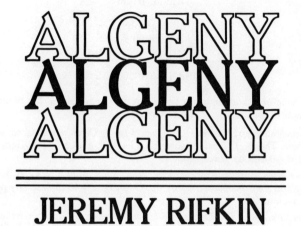

JEREMY RIFKIN

In collaboration with
Nicanor Perlas

THE VIKING PRESS   NEW YORK

LIBRARY OF CONGRESS CATALOGING IN PUBLICATION DATA
Rifkin, Jeremy.
Algeny.
Bibliography: p.
Includes index.
1. Genetic engineering—Social aspects.   I. Title.
QH442.R53   1983   303.4′83   82-40390
ISBN 0-670-10885-5

Grateful acknowledgment is made to the following for permission to reprint copyrighted material:

Academic Press, New York: Evolution of Living Organisms: Evidence for a New Theory of Transformation by Pierre P. Grassé. Copyright © 1977. Used by permission.

American Scientist: Review of *Implications of Evolution* by G. A. Kerkut in *American Scientist,* Vol. 49. Copyright © 1949. Used by permission.

Inquiry Press: The Creation-Evolution Controversy: Toward a Rational Solution by R. L. Wysong. Copyright © 1976 by Randy L. Wysong. Used by permission.

International Publishers, New York: Correspondence by Karl Marx and Friedrich Engels. Translated and edited by Dona Torr. Copyright 1934. Used by permission.

Lawrence & Wishart Ltd., London: Selected Works by Karl Marx and Friedrich Engels. Copyright 1943. Used by permission.

W. W. Norton & Co., Inc.: Darwin and the Darwinian Revolution by Gertrude Himmelfarb. Copyright © 1959, 1962 by Gertrude Kristol. *Victorian People and Ideas* by Richard D. Altick. Copyright © 1973. Used by permission.

Sierra Club Books: Nature's Economy: The Roots of Ecology by Donald Worster. Copyright © 1977. Used by permission.

University of Chicago Press: "Morphology, Paleontology, and Evolution" by Everett Claire Olson, in *Evolution After Darwin,* Vol. I, page 523, edited by Sol Tax. Copyright © 1960 by the University of Chicago. Used by permission.

Printed in U.S.A.
Set in CRT Baskerville

For Donna,
my love and my life

ACKNOWLEDGMENTS

I would like to give special thanks to John Gilbert Widrick for overseeing the production side of this project. Mr. Widrick has been responsible for guiding this manuscript through its various stages from research to copy editing. His dedication, commitment, and expertise have been invaluable. I would also like to thank Mr. Widrick for his many constructive editorial and conceptual suggestions.

My thanks also go to Donna Wulkan for her ongoing critique of the manuscript and for the countless hours of fruitful discussion in which many of the ideas of the book took form.

I would also like to thank Wes Michaelson for his editorial advice at a critical moment in the book's development.

For their patience and thoughtfulness throughout the long ordeal, I would like to thank Joshua and Anie.

I am also grateful for the help provided us by John Melegrito at Gelman Library, George Washington University.

Finally, I would like to thank my editor, Alan Williams, at Viking, whose critical evaluation and suggestions helped immensely in transforming this manuscript into its present form.

AUTHOR'S NOTE

While the nation has begun to turn its attention to the dangers of nuclear war, little or no debate has taken place over the emergence of an entirely new technology that in time could very well pose as serious a threat to the existence of life on this planet as the bomb itself. With the arrival of bioengineering, humanity approaches a crossroads in its own technological history. It will soon be possible to engineer and produce living systems by the same technological principles we now employ in our industrial processes. The wholesale engineering of life, in accordance with technological prerequisites, design specifications, and quality controls, raises fundamental questions. *Algeny* is a critique of the new technological epoch, the emerging concept of nature that will accompany it, and the intellectual foundation of Western thought that underlies it.

CONTENTS

PART ONE

FROM ALCHEMY TO ALGENY

A New Metaphor for the Coming Age

We are nearing the end of a unique period in human history. The Industrial Revolution, which began in the coalfields of England over two hundred years ago, is now entering its final stage in the oil fields of the Middle East. In this short span of history, we extracted hundreds of millions of years of stored sun and converted it into the utilities of modern life. In fact, the Industrial Age can be viewed as a convenient metaphor for the process of transforming, exchanging, and discarding nonrenewable energy. Our entire economic structure is built from and propelled by fossil fuels. We have invaded the long-silent burial grounds of the Carboniferous Age, appropriating the dead remains of yesteryear for the use of the living today.

Our buildings and factories, our machines and clothes and roads and vehicles, exist as a kind of ghoulish testimonial to our violation of the past. It is ironic indeed that we have called our age the Age of Growth. We are convinced that we have discovered a process for churning out perpetual wealth when instead we have only been borrowing from the past. As we begin to turn the corner on the Industrial Age, a nagging truth that has long festered beneath the surface of our collective consciousness begins to taunt the uneasy self-confidence that has characterized the modern age. We are, in truth, the wily and ingenious scavengers of history. We have fashioned great monuments from the corpses of the past, and now that we have exhumed their last remains, we find ourselves forced to ascend from the netherworld of stored sun back into a more familiar world bathed in sunlight.

The end of the age of fossil fuels presages the end of the Industrial Age that has been molded from it. This is the hard reality that is so difficult to fathom. The great Industrial Age is already passing from view. Its glass-and-steel monuments, once objects of wonder, are now worn and in disrepair. While still able to pique our curiosity, the marvels of the Industrial Age have become a matter of fact, if not indifference. At best, we have come to accept the accoutrements of the industrial era as we do the oxygen we breathe. At worst, we have become increasingly irritated with the aging process of the system itself, a fact that has forced us to pay greater attention to the many infirmities that seem to heap up, one on top of the other, with each succeeding day. Like an ancient relative that has become more of a burden than a blessing, the Industrial Age is now seen more in actuarial terms than in pioneering metaphors. In place of the idea of an invincible force that will usher in the material cornucopia, our policy leaders now speak of the industrial machine in tentative terms, their hopes limited to the narrow sphere of repair and maintenance. Gone is the once grand vision of the world paved in metallic sheen, with wheels racing noiselessly over its surface. Today we spend far less time planning the future and far more time eking out what is left of the past.

The industrial epoch marks the final stage of the age of fire. Its passage signals a great turning point for humankind. After ten thousand years of torching fire to ore, the age of pyrotechnology is slowly burning out. Even as we dip down into the last nooks and crannies deep beneath the earth's surface in search of remaining pockets of stored sun that can be extracted and used to keep the furnaces afire, we sense, though dimly, a new dawn emerging, one our eyes have never before fixed upon.

The age of pyrotechnology has been distinguished by the transition of the earth from a place people were exposed to, to a place people remade and surrounded themselves with. With the aid of fire, human beings turned the earth into an extension of themselves. Fire conditioned humanity's entire existence. In the *Protagoras* Plato recounts how human beings came to possess fire and the pyrotechnological arts. According to the myth, as the gods began the process of fashioning living creatures out of earth and fire, Epimetheus and Prometheus were called upon to provide them with their proper qualities. By the time they came to human beings, Prometheus noticed that Epimetheus had already distributed all the qualities at his disposal to the rest of the plants and animals. Not wanting to leave human beings totally unprotected, Prometheus stole the mechanical arts and fire from the gods and gave them to men and women. With these acquisitions, humanity acquired knowledge that originally belonged only to the gods. Prometheus' name means foresight, so it is only appropriate that this god be the one to aid humankind in its struggle to prevail against the forces of nature.

Fire, says Lewis Mumford, provided human beings with light, power, and heat—three basic things necessary for survival. Commenting on the role of fire in human development, Mumford concludes that it "counts as man's unique technological achievement: unparalleled in any other species."[1] With fire, human beings could melt down the inanimate world of nature and reshape it into a world of pure utilities. As historian Theodore Wertime of the Smithsonian Institution observes:

> ... there is almost nothing that is not brought to a finished state by means of fire. Fire takes this or that sand, and melts it, according to the locality, into glass, silver, cinnabar, lead of one kind or another, pigments or drugs. It is fire that smelts ore into copper, fire that produces iron and also tempers it, fire that purifies gold, fire that burns stone [cement] which causes the blocks in buildings to cohere.[2]

The age of pyrotechnology began in earnest around 3000 B.C. in the Mediterranean and Near East when people shifted from the exclusive use of muscle power to shape inanimate nature to the use of fire. Pounding, squeezing, breaking, mashing, and grinding began to play second fiddle to fusing, melting, soldering, forging, and burning. By refiring the cold remains of what was once a fireball itself, human beings began the process of recycling the crust of the planet into a new home for themselves.

Now that humanity has fashioned this second home, it finds itself in short supply of the raw energy and resources necessary to maintain that home. At the same time, the problem is compounded by the fact that this new home is increasingly inhospitable to the rest of life's creations, which are largely unable to adjust to this alien "manmade" environment. To put the magnitude of the problem in perspective, it is estimated that during the dinosaur age, animal species became extinct at a rate of about one per thousand years. By the early stages of the Industrial Age, animal species were dying out on the average of one per decade. Today, as we enter the final period of the long age of pyrotechnology, plant and animal species are dying off at the rate of one every sixty minutes. According to Thomas Lovejoy of the World Wildlife Fund, nearly 17 percent of all the plant and animal species remaining on earth will become extinct between now and the year 2000.[3]

Humankind, then, faces two crises simultaneously. The earth is running low on its stock of burnable energy and on the stock of living resources at the same time. We are at a turning point in the history of civilization, and it is at this critical juncture that a revolutionary new approach to organizing the planet is being advanced, an approach so overwhelming in scope that it will fundamentally alter humanity's entire relationship to the globe.

After thousands of years of engineering the cold remains of the earth into utilities, human beings are now setting out to engineer the internal biology of living organisms in the hope of staving off the crisis at hand and laying the base for a new world epoch. We are moving from the age of pyrotechnology to the age of biotechnology. The transition is indeed staggering.

For thousands of years humanity used fire to convert the earth's crust into new shapes and forms that never existed in nature. Now, for the first time in history, humanity has found a way to convert living material into new shapes and forms that never existed in nature. In 1973, American scientists performed a feat in the world of living matter to rival the importance of fire itself. Biologists Stanley Cohen of Stanford University and Herbert Boyer of the University of California reported that they had taken two unrelated organisms that would not mate in nature, isolated a piece of DNA from each, and then hooked the two pieces of genetic material together. The result was literally a new form of life, one that had never before existed on the face of the earth.

A product of nearly thirty years of investigation, climaxed by a series of rapid discoveries in the late 1960s and 1970s, recombinant DNA is a kind of biological sewing machine that can be used to stitch together the genetic fabric of unrelated organisms. Dr. Cohen divides recombinant DNA surgery into four stages. To begin with, a chemical scalpel, called a restriction enzyme, is used to split apart the DNA molecules from one source—a human, for example. Once the DNA has been cut into pieces, a small segment of genetic material—a gene, perhaps, or a few genes in length—is separated out. Next, the restriction enzyme is used to slice out a segment from the body of a plasmid, a short length of DNA found in bacteria. Both the piece of human DNA and the body of the plasmid develop "sticky ends" as a result of the slicing process. The ends of both segments of DNA are then hooked together, forming a genetic whole composed of material from the two original sources. Finally, the modified plasmid is used as a vector, or vehicle, to move the DNA into a host cell, usually a bacterium. Absorbing the plasmid, the bacterium proceeds to duplicate it endlessly, producing identical copies of the new chimera. These are called clones.

The recombinant DNA process is the most dramatic technological tool to date in the growing biotechnological arsenal. The biologist is learning how to manipulate, recombine, and reorganize living tissue into new forms and shapes, just as his craftsmen ancestors did by firing inanimate matter. The speed of the discoveries is truly phenomenal. It is estimated that biological knowledge is currently doubling every five years, and in the field of genetics, the quantity of information is doubling every twenty-four months. We are virtually hurling ourselves into the age of biotechnology.

Consider, for example, just ten of the scores of major advances made over the last two decades.

1. We have learned how DNA reproduces itself. In 1957, Stanford University biochemist Dr. Arthur Kornberg determined that the structure of DNA replicates by "unzipping" itself. Each strand then attracts new chemical substances from the surrounding cell, making an exact duplicate of its original structure. Since Kornberg's original revelation, we have learned how to reproduce copies of DNA artificially in a form clearer and more stable than the original.[4]

2. We have cracked the DNA code. The base units along the double helix form into what are called DNA triplets, a series of three-letter words. There are sixty-four possible combinations of these words, each triplet coding the instructions for production of a specific protein. Large quantities of proteins, in differing combinations, produce the variations in organisms. In 1961, Nobelist Dr. Marshall W. Nirenberg performed the biological equivalent of deciphering the Rosetta stone by isolating a DNA triplet and determining the protein it produced. Soon afterward, other biologists managed to analyze the sixty-four-word DNA code.[5]

3. We have learned how DNA transmits its instructions to the cell. Even with Watson and Crick's discovery of the structure of DNA, scientists still did not understand how a gene in a DNA molecule sends out its chemical instructions for the building of proteins. Then, in the early 1960s, several teams of researchers in the United States and France isolated messenger RNA, the mechanism that acts as the DNA information carrier.[6] Since that time, other forms of RNA have been found and their purpose determined. In April of 1977, a team of scientists at the University of California at San Francisco announced that they had successfully "ordered" messenger RNA to reproduce a strand of the original DNA that had dispatched it.[7]

4. We have analyzed chromosomes to determine genetic function. In 1956, researchers determined that the human cell carries forty-six chromosomes, packages of DNA on which the genes are located. A chromosome sample can now be taken from a three-month-old fetus, and through chromotagraphic analysis

the possible occurrence of some sixty genetic diseases can be predicted.[8]

5. We have synthesized a cell. Dr. James F. Danielli, professor of Theoretical Biology at Worcester Polytechnic Institute, headed the team that first built a cell. They accomplished the feat by reassembling parts of three different amoebas into one functional whole.[9] Scientists have also produced "giant cells" with the ability to grow five hundred to a thousand times larger than normal. Likewise, mini-cells have been engineered.[10]

6. We have fused cells from two different species. We can now take cells from two different organisms—for example, a mouse and a human—and fuse them together, producing a hybrid cell that carries some of the properties of the two originals. Cell fusion has also been accomplished between human and plant cells, hens' red blood cells and yeast, and carrot cells and cells taken from a human cancer victim.[11]

7. We have isolated pure human genes. In 1969, Dr. Jonathan Beckwith and a team of researchers at Harvard reported the first isolation of a pure gene from bacteria.[12] By 1973, the purifying of the first human gene had been accomplished. With the gene isolated from varied chemical reactions that take place in the cell, biologists can analyze it in a test tube. One outgrowth of this work is the increased understanding of the mechanism that turns genes off and on.[13]

8. We have "mapped" genes. The genes responsible for various physical traits like hair color are located at specific sites on specific chromosomes. Scientists are learning to locate or map genes on the chromosome. The mapping process began in the early 1970s; today, well over two hundred genes have been mapped.[14]

9. We have synthesized a gene. During the last decade, Dr. Har Gobind Khorana of M.I.T. pioneered in the synthesis of genetic material. In 1970, Khorana succeeded in synthesizing a gene found in a yeast cell. By 1976, he had performed a considerably more dazzling trick in building from scratch, using only the basic four nucleotides, a two-hundred-sequence human gene. As an added feature, Khorana equipped his creation with the start and stop mechanisms critical to controlling its function. Once

inserted into a cell, the synthetic gene proceeded to work perfectly.[15]

10. We have changed the heredity of a cell. In 1971, a team of
scientists introduced a gene from a bacterium into a human cell,
where it functioned, thus changing the original instructions received by the cell.[16] In another highly significant experiment in-
volving cell heredity, cells from one mouse were injected into the
fetuses of others. Not only did the genetic material from the first
mouse show up in the newborn mice, but when these mice
mated and had offspring, the foreign genes were passed on to the
next generation. Genetic alteration had been permanently incorporated into the germ line of a mammal, a development that
biologist Ethan Signer of M.I.T. believes "brings us ten years
closer to the possibility of genetic engineering of humans."[17]

While all these advances appear esoteric, and a bit opaque, their
full import is readily understood when applied to life-size organisms in the animal and plant kingdoms. In September 1981,
scientists working at the University of Ohio and in Jackson Laboratory in Bar Harbor, Maine, announced that they had successfully transferred a gene from one animal species to another.
Dr. Thomas Wagner, who headed the project, reported that they
had effectively isolated the gene that directed the manufacture
of part of the hemoglobin in rabbits. They then transferred the
rabbit gene into the fertilized egg of a mouse and brought the
fetus to term. The newborn mouse was unlike any other mouse
in history, for it not only contained the genetic material from a
rabbit, but the particular gene that was transferred produced
the hemoglobin product as effectively as it had in the rabbit.
The newly incorporated rabbit gene was then successfully passed
on through the normal mating process into subsequent generations of mice. It is now possible, the scientists concluded, to create new animal breeds by combining genetic information from
unrelated species.

Although spectacular in their own right, all these advances
represent just the first crude tools of the biotechnical age. Humanity is dramatically changing the way it goes about organiz-

ing the world. For thousands of years human beings have been fusing, melting, soldering, forging, and burning inanimate matter into economic utilities. Now they are about to begin the process of slicing, recombining, inserting, and stitching living material into economic utilities. The possibilities, say the scientists, are limited only by the span of the human imagination and the whims and caprices of the marketplace.

Lord Ritchie-Calder, the British science writer, cast the biological revolution in the proper historical perspective when he observed that "just as we have manipulated plastics and metals, we are now manufacturing living materials. . . ."[18] According to a recent study by the government's Office of Technology Assessment, bioengineering "can play a major role in improving the speed, efficiency, and productivity of . . . biological systems."[19] Our ultimate goal is to rival the growth curve of the Industrial Age by producing living material at a tempo far exceeding nature's own time frame and then converting that living material into an economic cornucopia.

In June 1980, the U.S. Supreme Court ruled that novel new forms of life engineered in the laboratory were patentable. In the aftermath of that historic decision, bioengineering technology shed its pristine academic garb and bounded into the marketplace, where it was heralded by many analysts as a scientific godsend, the long-awaited replacement to a dying industrial order. So anxious was Wall Street to begin financing the biotechnical revolution that when the first privately held genetic engineering firm offered its stock to public investors, a buying stampede within the investment community was nearly set off. On October 14, 1980, just months after the Supreme Court cleared the way for commercial exploitation of bioengineering technologies, Genentech offered over one million shares of stock at $35 per share. In the first twenty minutes of trading, the stock climbed to $89 a share—a gain of 54 points. By the time the trading bell had rung in late afternoon, the fledgling biotechnology firm had raised $36 million and was valued at $532 million. The astounding thing was that Genentech, now worth nearly one half billion dollars, had yet to introduce a single product into the marketplace. Commented one financial analyst from Merrill Lynch, "I have been with the firm twenty-two years [and] I have never seen anything like this."[20]

Genentech is one of a number of new bioengineering firms that are beginning to set the pace for the biotechnical revolution. With names like Biogen, Cetus, and Genex, these pioneers are blazing a trail for what some financial experts regard as the second great technological revolution in world history. According to Nelson Sneider, an investment analyst for E. F. Hutton, a firm that has nurtured much of the interest in the emerging biotechnical field, "We are sitting at the edge of a technological breakthrough that could be as important as . . . [the] discovery of fire."[21] Apparently many of the corporate giants are convinced. Dozens of the world's leading transnational corporations are pouring funds into biotechnical research. They include Du Pont, General Electric, Upjohn, Exxon, Allied Corporation, Phillips Petroleum, Kodak, and Dow Chemical.

In every business field, development guidelines are being laid out, long-range retooling of equipment is being hurried along, new personnel are being hired, all in a mad rush to introduce the life sciences into the economy, readying civilization to taste the first fruits of the biotechnological age. The sheer scope of the emerging technological revolution is awesome. Says E. F. Hutton's Nelson Sneider, "When you add up all the industries that could be impacted by biotechnology, you're dealing with up to 70 percent of the gross national product"[22] by the year 2010.

In the pharmaceutical industry, those in the know predict that bioengineering will revolutionize the production of antibiotics, enzymes, antibodies, vaccines, and hormones.

In the mining industry, there is experimentation going on with the development of new microorganisms that can replace the miner and his machine in the extraction of ores. Tests are being conducted with organisms that will eat metals like cobalt, iron, nickel, and manganese. One company reports that it has already successfully blown a certain bacterium "into low-grade copper ores where it produces an enzyme that eats away salts in the ore, leaving behind an almost pure form of copper."[23] For low-grade ores that are difficult to tap with conventional mining techniques, microorganisms will provide a more economical approach to extraction and processing.

In the energy industry, the oil companies are beginning to experiment with renewable resources as a substitute for oil and gas. Scientists hope to improve on existing crops, like sugar cane, which is already producing alcohol for automobile consumption. The future is likely to see the emergence of biologically engineered "fuel crops," whose sole function will be to produce usable energy for society.

In the chemical industry, scientists are talking about replacing petroleum, which for years has been the primary raw material for the production of chemicals, with biomass, a renewable resource made up of plant and animal material.

In agriculture, bioengineering is being looked to as a substitute for petrochemical farming. Scientists are busy at work engineering new food crops that can take in nitrogen directly from the air, rather than having to rely on the more costly petrochemical-based fertilizers presently in use. There are also efforts under way to increase the photosynthetic capability of selective plants in order to increase yield. In addition, experiments are under way to transfer desirable genetic characteristics from one species to another in order to increase productive performance. Scientists are trying to locate genes that help ward off viruses and pests, and that can adapt a plant to salty or dry terrains, all in an effort to upgrade the flow of living material into economic utilities.

In the field of animal husbandry, new bioengineering technologies allow man to bypass the slow, often unpredictable process of natural breeding. New genetic traits can now be programmed directly into the fetus. Nor are scientists any longer constrained by species boundaries. In the future the transfer of genetic characteristics between species will be commonplace. The goal will be to engineer new forms of animals that can meet specific economic demands. For example, Dr. Thomas Wagner suggests that it might be cost-effective to engineer cattle that can grow as well on grass or hay as on the more expensive feed grains. Since the buffalo is already adapted to a roughage cuisine, Wagner suggests that "you might transfer this trait alone of a buffalo to a cow, and leave all the rest of the buffalo behind."[24] Cloning offers still another possibility. Many scientists believe that by the end of the current decade, entire herds of domestic livestock will be cloned in order to ensure a uniform high-quality yield of meat and meat by-products.

Finally, there is the question of engineering the human anatomy. Many of the bioengineering techniques that prove successful in animals and plants can be adapted to some degree to the human frame. Scientists are already looking to the day when "harmful" genetic traits can be eliminated from the fetus at conception. Eliminating the specific genes that are the cause of many dreaded diseases only scratches the surface of the possibilities that lie ahead. Researchers believe that when today's babies are old enough to have children of their own, they may be able to select from a wide range of beneficial gene traits they would like to have programmed directly into their offspring at the fetal stage, from manual dexterity skills to improved memory retention capability.

Meanwhile, gene surgery is also likely to be a medical reality within a matter of a few years. Scientists predict that specially engineered genes will be introduced directly into the human body in order to produce agents that will immunize against specific diseases. Other genes will be inserted that can help facilitate or retard growth, regenerate limbs, and perform a host of other medically useful activities.

The thought of recombining living material into an infinite number of new combinations is so extraordinary that the human imagination is barely able to grasp the immensity of the transition at hand. These first few processes and products are the biotechnical equivalent of the first pots and bins forged by our ancestors tens of thousands of years ago when they began experimenting with the pyrotechnical arts for the first time. From the moment our Neolithic kin first fired up the earth's material, transforming it into new forms, humanity locked itself into an irreversible journey that finally culminated in the Industrial Age. Our world, the world of twentieth-century man, is forged in fire. Its skyscrapers and satellites and sewers and electrical lines and highways and homes and virtually every other economic convenience are the final fruits of the pyrotechnical revolution begun eons ago when a few enterprising ancestors decided to torch a piece of ore with a hot flame.

Now humanity has set its sights on the living world, determined to reshape it into new combinations, and the far-distant consequences of this new journey are as unfathomable to today's biotechnologists as the specter of industrial society would have been to the first pyrotechnologists.

Accompanying this great technological transformation is a philosophical transformation of monumental proportions. Humanity is about to fundamentally reshape its view of existence to coincide with its new organizational relationship with the earth.

The best way to understand this conceptual revolution is by comparing two metaphors. For the age of pyrotechnology alchemy serves as a convenient conceptual metaphor. For the age of biotechnology the appropriate conceptual metaphor is algeny.

When we think of the term "alchemy" today, what immediately comes to mind is the futile search for a method by which lead could be transformed into gold. As Morris Berman points out in his book *The Reenchantment of the World*, it was much more. According to Berman, alchemy was at one and the same time "the science of matter, the attempt to unravel nature's secrets; a set of procedures which were employed in mining, dyeing, glass manufacture, and the preparation of medicines; and simultaneously a type of yoga, a science of psychic transformation."[25]

Alchemy is said to have originated as a formal philosophy and process in Egypt during the fourth century B.C., although many historians believe that its roots lie much further in antiquity, dating as far back as the first city-states at Sumer.

According to the alchemist, "All metals are in the process of becoming gold."[26] They are, in other words, gold *in potentia*. The alchemists believed that every metal is continually seeking to transform itself, to transcend its original state and experience its true nature, which they said is gold. Meister Eckhart, the thirteenth-century mystic, once remarked that "copper is restless until it becomes gold."[27] The alchemists were firmly convinced that it was possible to accelerate what they believed to be a "natural process" of transformation by way of an elaborately orchestrated set of laboratory procedures.

The alchemic process began with the fusing together of several metals into one mass or alloy, which was then regarded as a kind of universal base material from which the various transmutations could be made. Fire, of course, was indispensable to the entire transmutational process. It allowed the alchemist to melt, fuse, purify, putrify, distill, and coagulate his base material, creating new combinations and forms, each one closer to the ideal golden state.

Alchemy, then, was a philosophy and a technical activity at the same time. Nature, for the alchemist, was a process attempting to complete itself. The alchemist, as Morris Berman points out, viewed himself as a midwife, "accelerating" the natural process, hurrying the physical world along to its own perfected state. Alchemy is believed to have been a derivative of an Arabic word meaning "perfection." Humanity's task, according to the alchemists, is to help nature in its struggle to "perfect itself." Gold, in its seeming permanence and in its strong resilience to fire, conjured up the image of immortality and perfection.

Alchemy elevated the pyrotechnical arts to a sacred status. Using fire to heat and transform materials from one state to another became associated with a higher purpose: the completion and perfection of the natural process.

Anthropologists believe that the alchemic tradition grew out of the arts of fabric dyeing and of bronzing or coloring of metals, which reached its zenith in ancient Alexandria. The precursors of the alchemists were artisans attempting to find cheap metal substitutes for gold. What began as art imitating nature eventually became transformed by the late Middle Ages into an all-embracing explanation of the workings, purpose, and goal of the natural order itself. So convinced were the alchemists that what they were doing in the laboratory was an integral part of the natural process that they came to believe that the "gold" they created was not really an imitation at all but rather a superior form of gold, one that represented the perfect state to which all natural gold aspired. Eventually, alchemy as a philosophy and as a process evolved into the modern scientific world view, which, to this day, clings to the notion of transforming and perfecting nature by imitative procedure.

Now, as we move from a pyrotechnical to a biotechnical relationship with nature, a new conceptual metaphor is emerging. Algeny is about to give definition and purpose to the age of biotechnology. The term was first coined by Dr. Joshua Lederberg, the Nobel laureate biologist who now serves as president of Rockefeller University. Algeny means to change the essence of a living thing by transforming it from one state to another; more specifically, the upgrading of existing organisms and the design of wholly new ones with the intent of "perfecting" their performance. But algeny is much more. It is humanity's attempt to give metaphysical meaning to its emerging technological relationship with nature. Algeny is a way of thinking about nature, and it is this new way of thinking that sets the frame for the unfolding of the next great epoch in history.

An algenist views the living world as *in potentia*. Life is seen as a process in which every organism is seeking to complete itself. In this regard, the algenist doesn't think of an organism as a discrete entity but rather as a temporary set of relationships existing in a temporary condition, on the way to becoming something else. For the algenist, species boundaries are just convenient labels for identifying a familiar biological condition or relationship, but are in no way regarded as impenetrable walls separating various plants and animals. The algenist contends that all living things are reducible to a base biological material, DNA, which can be extracted, manipulated, organized, combined, and programmed into an infinite number of combinations by a series of elaborate laboratory procedures. By engineering biological material, the algenist can create "imitations" of existing biological organisms that to his mind are of a superior nature to the originals being copied.

The final goal of the algenist is to engineer the perfect organism. The "golden state" is the state of optimal efficiency. Nature is seen as a hierarchical order of increasingly efficient living systems. The algenist is the ultimate engineer. His task is to "accelerate" the natural process by programming new creations that are more "efficient" than those that exist in the state of nature.

Algeny is both philosophy and process. It is a way of perceiving nature and a way of acting on nature at the same time. It is a revolution in thought commensurate in scale to the revolution in technology that is emerging. We are moving from the alchemic metaphor to the algenic metaphor.

Up to this point in history, humanity's ability to manipulate living things has always been dependent on its ability to alter their environments. People could decide where plants, animals, and other human beings were to live, what they were to eat and how much, and even whom they were to mate with. People could domesticate living things, including themselves, and they could even vary looks, weight, and other secondary characteristics through careful breeding. What they couldn't do was either fundamentally change the existing structure of plants, animals, and people or create wholly new structures. Human beings have had to live with the constraints imposed by existing biological forms. People could cultivate but they could not engineer or create.

In all of humanity's past experience, living things enjoyed a separate, unique, and identifiable place in the order of nature. There were always rabbits and robins, oaks and ostriches, and while human beings could tinker with the surface of each, they couldn't penetrate to the interior of any. Now, as we move from the age of pyrotechnology to the age of biotechnology, people are beginning to learn how to reorganize living things from the inside out. The redesign of existing organisms and the engineering of wholly new ones mark a qualitative break with humanity's entire past relationship to the living world. People's reconception of nature is going to change just as radically as their organization of it.

Engineering new forms of life requires a wholesale transformation of our thought patterns. It should be remembered that the entire way we formulate our conception of the world is etched in the fires of the age of pyrotechnology. We are dousing the Promethean flame, and the new world we are entering is alien to the vision of all the great theologians, philosophers, and metaphysicians of the past. As we move from firing dead ores to penetrating living tissues, as we invade the interior of living organisms with engineering tools, as we begin to plot new designs for the reconstruction of life itself, the voices of past seers fall silent. Their words were never meant to extend to this new epoch, were never meant to explain this emergent reality.

Already, a great schism is developing between the last generation of the age of pyrotechnology and the first generation of the age of biotechnology. Our children are beginning to conceptualize the world in a fashion so fundamentally different from anything we can readily identify with that the empathetic association that traditionally passes down through the generations, uniting past with future, seems at times to be irretrievably severed—as if to suggest the termination of one great lifeline in history and the abrupt beginning of another. Our children are the first sojourners of the second great economic epoch. While they still carry with them most of the conceptual trappings of the age of fire, they are beginning to experience the world from a profoundly altered frame of reference.

To begin with, their language is the language of the computer. Their world of communications is made up of computer programs, electronic games, word processors, videodiscs. The average American child now spends approximately twenty-eight hours per week with electronic learning tools, compared to twenty-five hours per week with printed learning materials. The electronic image and the computer printout are increasingly taking the place of the spoken and written word. A *New York Times* article reports mathematician Seymour Papert of M.I.T. as saying that "the effect of the computer on learning and thinking is comparable to that of the invention of writing."[28]

Alan Newall of Carnegie-Mellon University is one of the experts in the new field of artificial intelligence and the computer sciences. He argues that the true import of the computer is that it opens students' minds to a "whole new language for describing behavior."[29] This new language is drastically altering our children's perception of the world. Many educators now believe that our young people are beginning to conceptualize the world in the same terms that animate the operations of a computer system.

Joseph Weizenbaum, professor of computer science at M.I.T., best expressed the conceptual revolution that separates the generations when he observed: "To him who has only a computer, the world looks like a computer domain."[30] The world as seen from a computer perspective is very different from the one we have experienced in the past. In this new world, all physical phenomena are reduced, reorganized, and redefined to meet the operating requirements of the computer. The computer recasts the world in its own image, transforming all of nature into bits of information to be processed and programmed. In point of fact, the computer creates a new context for organizing the world, one that supersedes the industrial frame. It is this new "context" that the first generation of computer babies is being prepared for.

Most futurists have yet to perceive the full significance of the computer revolution. While they have been correct in their appraisal of the impact of the computer and the information sciences, they have misunderstood their ultimate role in the future of civilization. Most futurists see the computer revolution as a new method for organizing the Industrial Age. While it is true that the computer has been successfully adapted to industrial processes, it should be noted that its appearance coincides with the final stages of the industrial era. The Industrial Age peaked in the early 1970s. Ever since that time, the world community has been finding it more and more difficult to locate and process a dwindling supply of nonrenewable energy. Because nonrenewable energy is the organizing material of the Industrial Age, its depletion marks the end of the economic era built from it and maintained by it.

The entire industrial era ran its course without the aid of the computer. This new organizing mechanism didn't come "on line" until the mid-1960s. It didn't begin to exert a commanding presence until the early 1980s. The computer only caught the tail end of the industrial era. While it will, no doubt, be used in a myriad of ways to stretch out the remaining years of the industrial epoch, its real import has yet to be gleaned by the future forecasters.

The computer is the organizing mechanism for the age of biotechnology, just as the industrial machine was the organizing mechanism for the Industrial Revolution. Whereas the machine transformed nonrenewable resources into economic utilities, the computer will transform biological material into economic products and processes.

The computer is also the language of the biotechnical age. Every great economic period brings with it a unique form of communication. Hunter-gatherer societies relied on sign and oral language, while every advanced agricultural society had some form of written language. The printing press was used during the early stages of the Industrial Revolution. No self-respecting anthropologist, however, would refer to the Paleolithic period as an oral economy or the Neolithic period as a written economy or the Industrial Age as a print economy. Yet today's futurists believe that what lies ahead is the computerized information economy. They fail to understand that the computer and information sciences are not in and of themselves the new economy. Rather, they are the organizing language for the new economy. They are the means of communication that humankind will use to reorder living material in the biotechnical age.

In 1981, the first computerized gene machine made its debut. One need only type out the genetic code for a particular gene on the computer's keyboard and within a matter of a few hours "the machine delivers a quantity of synthetic gene fragments that can be spliced together and put into the DNA of living organisms."[31] With the gene machine it is possible to begin transforming living material into new designs and products in large enough volume and with sufficient speed to provide a cost-effective starting point for the biotechnical economy. This, however, is only the beginning stage of the coming economic revolution. Eventually scientists hope to mesh living material and the computer into a single mode of production. Already, corporate funds are being channeled into research designed to replace the microchip with the biochip and the microcomputer with the biocomputer. According to James McAlear, president of EMV, one of the firms pioneering in this research:

> Our aim is to build a computer that can design and assemble itself by using the same mechanism common to all living things. This mechanism is the coding of genetic information in the self-replicating DNA double helix and the translation of this chemical code into the structure of protein.[32]

Within the coming decade, the computer industry and the life sciences are expected to join together in a new field, molecular electronics. Companies like Japan's Mitsui Corporation are already planning for that day by acquiring "a large stake in both biotechnology and microelectronics."[33] The grand objective is to turn living material into biocomputers and then to use these biocomputers to further engineer living materials. In the future, biocomputers will be engineered directly into living systems, just as microcomputers are engineered into mechanical systems today. They will monitor activity, adjust performance, speed up and slow down metabolic activity, transform living material into products, and perform a host of other supervisory functions. Scientists even envision the day when computers made out of living material will automatically reproduce themselves, finally blurring the last remaining distinction between living and mechanical processes.

According to Dr. Zsolt Harsanyi, vice-president of DNA Science, the day of the biocomputer is within grasp. He predicts that by the time today's babies reach adulthood, the biocomputer could be a commonplace. The biocomputer represents the ultimate expression of the biotechnical age. By successfully engineering living material into an organic computer that can think, reproduce itself, and transform other living material into economic utilities, humanity becomes the architect of life itself in the coming age.

Although the biocomputer is not yet a reality, our children are already being prepared for the day when living systems will be programmed by computer design. As mentioned, the next generation is being immersed in the world of the computer. The computer is becoming so integral to every facet of our children's lives that they are coming to regard their whole environment as a computable domain. Once our children are comfortable with the idea of thinking of nature as "systems of information," they are all but ready for the task of programming nature by computer design.

For our children, then, nature will no longer be something they are born into but rather something they program. Already scientists are feverishly at work in the new field of computer graphics, attempting to bring this new biotechnical process "on line." In their book *Life for Sale,* science writers Sharon and Kathleen McAuliffe speculate that eventually

> computer graphics may be coupled with recombinant DNA technology, facilitating the production of genetic blueprints that correspond to make-believe molecules. In this way, pictures on a video display screen could readily be transformed into their real-life counterparts.[34]

With computer programming of living systems, the very idea of nature being made up of discrete species of living things, each with its own inviolate identity, becomes a thing of the past, a relic of the pre-biotechnical era. Simply by punching in the instructions on a keyboard, it will be possible to cross species walls and program an entire array of novel organisms. Our heirs will live in a world engineered and populated by their own creations. Imagine, then, how completely alien their perception of nature will be from our own and that of all those who preceded us. The order of difference is of a magnitude far greater than any that has ever separated one generation from another in world history.

The fact is, the next generation is going to entertain a radically new image of nature and humanity's relationship to it, one compatible with the bioengineering of living systems by computer design. If we are to understand the dimensions of the coming age, we need to begin with an understanding of the new concept of nature that is going to emerge among the first generation versed in the language of the computer and the biotechnical arts.

Every age has its own unique view of nature, its own interpretation of what the world is all about. Knowing a civilization's concept of nature is tantamount to knowing how a civilization thinks and acts. For our century, Darwin's theory of evolution has served as the centerpiece of the cosmological order. Five generations of human beings have accepted Darwin's interpretation of how nature works. We think of the world in a Darwinian way. We act in the world in a Darwinian manner. If we want to answer the ultimate futurist question of how generations not yet born will think and act differently from us, we need to know about the new concept of nature our children will adopt and how it will differ from our own.

This, then, is the story of the new view of nature that's going to replace Darwin's theory of evolution and provide a framework for the first generation of the biotechnical age. It's also the story behind the story; an account of the process by which the new concept of nature has been arrived at. To understand that process, we have to begin with an examination of the role concepts of nature play in the life of a civilization.

PART TWO

DECIDING WHAT'S NATURAL

The Ultimate Intellectual Deception

Concepts of nature always focus on the big questions: Where did we come from? Why are we here? Where are we headed? It's been that way from the time our ancestors took their first tentative stroll out of the forest to bask in the warm sunlight of the open savannahs. Of course, we don't dwell on these questions from dawn to dusk, during our every waking moment. For the most part, we humans go about our daily business with little regard for these larger questions. There are moments, to be sure, when our routine is rudely interrupted. It might be in the middle of a restless night of tossing and turning, our mind drifting in and out of consciousness, straying into areas best left unexplored; or it might be in the aftermath of an emergency, just as the first rush of adrenaline has begun to trail away, leaving us exposed and disoriented. Or it could be during one of those many times in life when we find ourselves face to face with a macabre experience that simply can't be explained, rationalized, or accounted for in a satisfactory way. These times are the most troubling in life. They force us to ask the primordial questions that have plagued humanity since the beginning.

Fortunately for our sense of personal well-being, life is filled
only partially with questions. For the most part, we live with answers, and it is here that any reappraisal of what life and existence are all about must begin. For as long as we have had a
history, human beings have had, at their disposal, a set of readily
available answers as to what nature and life are all about. Where
do these answers come from? How reliable are they? Why do answers we have long assumed to be beyond reproach suddenly
become objects of ridicule and scorn? Are the new answers that
replace them any more valid or are they doomed to the same ultimate ignominious fate?

The fact is, we human beings cannot live without some
agreed-upon idea of what nature and life are all about. When we
ponder what our own personal fate might be after the last breath
of life is extracted, or when we try to imagine what existed before existence itself, we are likely to become paralyzed with
doubt. Our concept of nature allows us to overcome these ultimate anxieties. It provides us with some of the answers, enough
to get along. A concept of nature, then, is more than just an explanation of how living things interact with one another. It also
serves as a reference point for deciphering the meaning of existence itself.

Concepts of nature have long been humanity's window to the
universe. From time immemorial human beings have been peer-
ing into the inner workings of nature in hopes of shedding light
on the secrets that govern the whole of existence. The assumption has always been that nature and the universe were interrelated in one unified design. It's no wonder, then, that for most of
recorded history humanity made little distinction between the
workings of nature and the workings of the cosmos. Concepts of
nature and cosmologies are both concerned with the meaning
and structure of existence, albeit on different planes. The separation of one from the other only began with the Greeks. Today,
scientists talk of cosmology and nature as totally separate phenomena. There are the laws of the universe and then there are
the laws of nature.

While the scientific community has erected a wall to separate
the two, philosophers still provide an arena where both terms

can intermingle. In philosophical discussions concerning the hows and whys of existence, it is not uncommon to hear the terms "cosmology" and "concept of nature" used interchangeably. Even here, however, the terms carry slightly different weights. A cosmology is a concept of nature that has been elevated and sanctified. When we talk about a concept of nature as a cosmological formulation, what we mean is that it provides a universal framework for interpreting existence. In this sense, cosmology is being used to express the idea of a world view or, as the Germans would say, a *Weltanschauung*.

It is rare to meet someone who questions Darwin's cosmology. It seems to make a great deal of sense when you stop to think about it. It answers, in a rather concrete way, some of the big questions. According to Darwin, everything evolved by chance, beginning with chemical mutations that combined by accident to become the first functioning life forms. Through repeated chance mutations and reorganizations over eons of time, life evolved, finally culminating in the most complex and interesting of all life's creations—ourselves. If it is difficult to find someone or some force to thank for the entire process or to properly assess where the process is eventually going to end up, at least it is comforting to know how the mechanism of life operates. With evolution we can feel reasonably safe about the world around us. When we gaze into our lover's eyes, touch a tree in the park, pet our family dog, or discover a spider's web tucked into the corner of our cedar closet, a cold, alien chill doesn't shoot up our spine, as it would if we were suddenly swept up into an alien universe buried in a constellation of stars somewhere in the farthest recesses of space. The reason we don't get hysterical is that we have a fairly good idea of how these things originated, what their purpose is, and, most important, what their relationship to us is in the total scheme of things. In short, we have a concept of nature that we believe in. That's not to say that we don't have some reservations about evolution. Certainly we're willing to admit that particular aspects of the theory are open to revision. But when we get right down to it, our belief, if not fervent, remains unshaken.

The litmus test for belief in evolution is simple. When we try to imagine an alternative explanation for the origin and development of life, we draw a complete blank. Our mind wanders into a dark hole where there exist not even a few measly strands that can be woven by the imagination into an alternative explanation. That's how we know evolution is the correct explanation of life. If there were a better explanation, certainly we would be able to fathom it.

Many different explanations for the creation and existence of nature and life have been extolled throughout history. It's interesting to note that in each case the prevailing view of a civilization or an epoch was adhered to with the same unshakable conviction with which we seem to hold to the theory of evolution. If asked to imagine an alternative concept of nature, our ancestors would likely have been just as incredulous of entertaining such a heresy as we are today. So, was everyone before Darwin wrong? Were all those countless conceptions of nature to be explained away as momentous human indiscretions born of ignorance and illusion? Is it possible that only one Charles R. Darwin, born in Shrewsbury, England, at the turn of the last century, was made privy to the great secret of life itself? Is humanity to be eternally grateful to this gentle Englishman for finally opening the door to the truth of existence? Have we finally turned the corner on the many errors and blind alleys that were pursued by humanity in its quest over the millennia for the how and why of life?

Let's back up a moment and consider a remark attributed to the distinguished philosopher of science C. D. Broad, of Cambridge University. "At certain periods in the development of human knowledge," he remarked, "it may be profitable and even essential for generations . . . to act on a theory which is philosophically quite ridiculous."[1] The key is the word "profitable," and Broad is using the word in more than just a pecuniary sense. Theories are useful. If they weren't we wouldn't find any need for them. They are useful precisely because they help in some way to facilitate the day-to-day reality we're engaged in. If there were no direct relationship between a theory and our day-to-day experience, the theory would not enjoy a very long life. But here's where the problem begins. We know that our day-to-day activity changes, sometimes qualitatively. The way we process the environment, the patterns we establish in order to live with each other, the kinds of mechanisms we fashion to organize our economic survival, are stable for only short periods of time. History is the process of continual change in the environment and in our relationship to it. A theory that has proven useful within a given setting may no longer be very helpful when the old way of doing things is replaced. Now, this might seem rather obvious when we are talking about political, social, or even economic theory, but what about when we cross the border into natural philosophy? Are concepts of nature purely instrumental and merely useful in some way to the day-to-day circumstances that people are engaged in? If so, then we are no longer justified in talking about evolution or any other concept of nature as iron-clad truths. Tools, perhaps, but not truths.

Our academic community has been rather schizophrenic when it comes to the question of scientific truths. It has long been accepted that "specialized bodies of thought and knowledge, such as aesthetic, moral and philosophical systems, religious creeds and political principles, are influenced by the social and cultural contexts in which they are produced."[2] No one seems to have a problem up to this point. It's when some scholars dare to suggest that scientific assumptions, theories, truths, and principles are also culturally influenced that the gloves come off, tempers rise, and blood begins to spill.

For, if our science is influenced by time, place, and circumstance, then we're indeed in big trouble. What, then, do we hold on to? How do we discriminate? What happens to our beliefs, our way of life? The prospect of an ultimate and irreconcilable uncertainty about the way the world operates is too overwhelming for the human psyche. So we continue to view our changing knowledge of nature as revelatory rather than relationary. The prevailing consensus is that there is an objective world out there that can be discovered by observation and that history shows a gradual, step-by-step accumulation of more and more accurate knowledge about its origin, development, and teleology.

According to Michael Mulkay, an authority on the sociology
of science at York University, England, most scholars have

> repeatedly rejected in principle the possibility that the form or
> content of scientific knowledge, as distinct from its incidence or
> reception, might be in some way socially contingent. Instead, they
> have argued strongly . . . that the substance of scientific knowl-
> edge is independent of social influence and they have tried to jus-
> tify this assertion on philosophical grounds.[3]

This traditional view that knowledge and truth about nature are
independent of context is now being challenged by a new generation of scholars. They are eager to expose this last sanctuary
of intellectual purity to the rigors of social reconstruction, and
their efforts, if successful, portend a profound shift in our approach to the big questions that animate the human discourse.

Otto Rank, one of the great psychoanalysts of the twentieth
century, suggests that our concepts of nature are supremely
anthropomorphic, reflecting our desire to make everything conform to our current image of ourselves. Rank believes that our
concepts of nature tell us more about ourselves at any given moment of time than they do about nature itself. Historian of science Robert Young of Cambridge University would agree with
Rank. He argues that there is no neutral naturalism. When we
penetrate to the core of our scientific beliefs, says Young, we find
that they are as much influenced by the culture as all our other
belief systems. More to the point, anthropologist C. R. Hallpike
of Dalhousie University in Canada contends that "the kinds of
representation of nature . . . that we construct . . ." flow from the
way "we interact with the physical environment and our fellows."[4] In a word, our concepts of nature are utterly, un-
abashedly, almost embarrassingly anthropocentric.

Try to imagine a society faithfully adhering to a concept of
nature that is at odds with the way it goes about structuring its
day-to-day activity. Obviously, a concept of nature must be
compatible with the way people behave within a given cultural
milieu if it is to be acceptable. This has always been the case. For
example, consider the cosmological foundation of one of the very

first civilizations in the world—the Sumerian city-states. In order to understand the Sumerian concept of nature, we need to look at how the Sumerians "interacted with their physical environment."

Sumer was the first of the large-scale hydraulic civilizations. Canals were dug, dikes were erected, and dams were built, taming the mighty floodwaters of the Tigris and Euphrates and directing their flow into the low-lying lands. The Sumerian system of irrigation was a technological tour de force—complex, delicate, and exceedingly difficult to maintain. It bathed the dry land with a gentle liquid nourishment, but at a cost. It required an elite group of managers to supervise its engineering and maintenance, and antlike armies of peasants and slaves to build, patch, and mend the mile upon mile of construction. The massive public works system brought with it elaborate political hierarchies, bloated government bureaucracies, privileged minorities, and exploited majorities. Written language was invented, laws were written down, and the art of political coercion and mass repression was practiced on a large scale for the first time in history. At Sumer, civilization begins to flow.

One word accurately sums up the relationship between the Sumerian concept of nature and the Sumerian social order: tautological. In this respect, the Sumerian experience is similar to the experience of every other major civilization. The Sumerian cosmology starts with the primeval sea. Interestingly enough, the Sumerians saw the sea as "a kind of first cause and prime mover,"[5] and Samuel Noah Kramer, professor of Oriental Studies at the University of Pennsylvania, tells us that they never asked the question of what came before the water. From the sea, the universe was generated, just as Sumer was generated from the irrigating floodwaters. The universe was ruled by a pantheon of gods, each performing some particular function in a kind of grand cosmic hierarchy. The entire operation was run in a manner that Kramer says was remarkably similar to the political operation of the Sumerian city-states, with a working assembly of fifty lesser deities headed by a smaller cadre of seven gods and a titular head at the apex of the structure. Kramer concluded that the Sumerian theologian "took his cue from human society as he knew it and reasoned from the known to the unknown."[6]

In the Sumerian language the word for water is also the word for semen. Enki, the water god, is also the god responsible for "generating" life. According to Sumerian myth, the land is "literally awash with semen."[7] Cultural historian William Irwin Thompson recounts an ancient Sumerian ode to "The Great Father" in which water overflows its boundaries, inseminating everything in its path. "The landscape of Sumer," says Thompson, "with its marshes, dikes and canals, is, therefore, a male landscape of irrigation technology, military societal organization, and male fertilizing power."[8] This conception of creation is light-years away from the Neolithic cosmologies in which all life is generated by the Great Mother. In Sumer, Mother Earth is passive matter, made alive only by the powerful generating ability of Enki, the male god.

Among the things Enki did to ensure the earth's "fertility and productiveness" was to fill the Tigris with fresh, clean water. He then tamed the current by mating with the water. Having created an elaborate irrigation system, he even appointed a minor god, Enbilulu, as "canal inspector," to make sure everything ran properly. What more could a Sumerian ask of his gods?

As Sumer so well illustrates, a society's concept of nature always turns out to be quite congenial with the way a society organizes itself and its environment. It could hardly be otherwise. To believe that we are instead merely neutral observers of the external world, and that those of us who are the most neutral are sometimes fortunate enough to stumble across one of nature's hidden secrets, is to believe in what environmental psychologist Bernard Kaplan referred to as "the dogma of immaculate perception."[9]

Rationalizations

Hunter-gatherers did it. Neolithic farmers did it. The Greek politicians did it. The Christian monks did it. The twentieth-century entrepreneur does it. Study any society anywhere at any time in history and you will find the same relentless need to inflate day-to-day activity into universal principles of nature. Why do we do it? What exactly compels us to enshrine the mundane banalities of life into timeless cosmic truths? There are countless reasons, many of which will be explored in other sections of the book. For now, two deserve some immediate attention.

By elevating the commonplace to the sublime, a society is able to provide a cosmic rationale to legitimize its economic, political, and social activities and at the same time eschew all responsibility for those activities.

Every civilization justifies its behavior by claiming to have the natural order on its side. In each case, the process of legitimization is the same. First, a society organizes itself and its environment. Hierarchies are set up. Relationships are determined. Tasks are allocated. Rewards are distributed. But how do the members of society know that the way they're going about their daily routine is the right way? This is the ultimate political question that every society faces. The political answer amounts to a conjurer's sleight of hand. Since a society's view of what the whole world is all about is heavily influenced by the way it is organizing its own immediate world each day, it is only "natural" for the culture to come to the conclusion that the economic, political, and social reality it feels and experiences must, in fact, be reality. Therefore, it is only a short jump to fashioning a model of nature that is strikingly similar to the world being fashioned by the society. Then, not surprisingly, people find that their behavior does indeed correspond to the order of nature and, for that reason, conclude that the existing social order is the right one, the correct one, the only one, and not to be trifled with. What better legitimization can there be for any governing body? Individuals rule and institutions prevail as long as enough people remain convinced that such behavior is merely a reflection of "the natural order of things." Concepts of nature are largely esoteric politics.

Indeed, concepts of nature may well serve as an essential political instrument for eliciting unequivocal "deference and resignation." No one in his right mind would suggest that it is correct or even possible to resist the natural order. And if society happens to be unjust, exploitive, repressive, what is a person to do? If it's merely a reflection of the natural order of things, or at least structured in a way that adheres to nature's grand design, then to challenge it in any fundamental way would be as foolhardy and self-defeating as challenging nature itself.

For society at large, and for ruling elites in particular, a concept of nature provides a double-edged sword. At the same time that it establishes a mantle of legitimacy for the existing social order, the prevailing concept of nature also allows the rulers to avoid ultimate responsibility for their behavior. Legitimacy without responsibility is the ultimate dream of every political elite, and it is through sanctification of a concept of nature that the feat is realized.

Perhaps the best contemporary illustration of how the process of "cosmological justification" works can be seen in the public discussion surrounding the nuclear bomb. In countless debates over the pros and cons of nuclear weaponry, invariably someone in the crowd reminds his fellows that the question is not whether we should have the bomb or not, but rather, how to manage its use. The popular reasoning goes something like this: If scientists are able to invent the atomic bomb and even more lethal weaponry, there is no way to stop that from happening. Nor should we try. After all, the very fact that they were able to imagine the weapon and then construct it is evidence of "evolution at work." According to the argument, the human mind had obviously "evolved" to the point that it could actually construct a nuclear bomb, and therefore the bomb itself is a product of evolutionary development. To attempt then to say no to its birth is more than just an assault on the sacred notion of freedom of scientific inquiry or the idea of progress. On a more basic level, resistance implies a challenge to the laws of nature, in this case, "the laws of evolution," which we believe to be inviolable and irreversible. How many times have we heard it said that it's impossible to put the genie back in the bottle once it has escaped. If we can imagine it and if we can realize it, then whatever the "it" is, its right to exist will prevail because, after all, it is a product of evolution.

The theory of evolution, then, provides a convenient rationale for avoiding responsibility for our activities.

Of course, this isn't a new story. Lest we adorn our current conception of nature with a bit too much cunning, it should be pointed out that this kind of cosmological justification has repeated itself with every concept of nature that humankind has ever promulgated.

For example, in the late medieval era, most Europeans accepted the official Church view of the origin of species spelled out by the great medieval schoolman St. Thomas Aquinas in the thirteenth century. St. Thomas borrowed heavily from Hebraic and Hellenic thought, adding some of his own ideas in the process. The result was a cosmology that legitimized the existing social order while relieving the powers that be of any responsibility for their behavior.

To begin with, Aquinas asked why the created order resembled a Great Chain containing a myriad of animals and plants in a descending hierarchy of importance. St. Thomas concluded that the proper workings of nature depended on the labyrinth of relationships, obligations, and dependencies among God's creatures. Geographer Clarence J. Glacken of the University of California says that as far as St. Thomas was concerned, God had intended that nature be populated with "many creatures differing among themselves in gradation of intellect, in form, and in species."[10] According to Aquinas, "Diversity and inequality"[11] guaranteed the orderly working of the system as a whole. The Churchman reasoned that if all creatures were equal, they could not "act for the advantage of another."[12] By making each creature different, God established a hierarchy of obligations and mutual dependencies in nature.

St. Thomas's characterization of nature bears a striking likeness to the institutional arrangement in medieval Europe, where there existed a tightly defined social structure in which everyone's individual survival depended on the dutiful performance of a complex set of mutual obligations within a rigidly maintained hierarchical setting. From serf to knight, from knight to lord, and from lord to Pope, all were unequal in degree and kind, each was obligated to the other by the medieval bonds of homage, and all together made up a mirror in which could be viewed, though only hazily, the perfection represented in God's total creation. According to historian Robert Hoyt of the University of Minnesota:

> The basic idea that the created universe was a hierarchy, in which all created beings were assigned a proper rank and station, was congenial with the feudal notion of status within the feudal hierarchy, where every member had his proper rank with its attendant rights and duties.[13]

In a related manner, there is also the question of determining one's proper place in the natural order. According to St. Thomas, "It would be inconsistent with the rationality of the divine government not to allow creatures to act according to the mode of their several natures."[14] It was Aquinas's belief that everything had been designed to perform in exactly the manner God intended, and for that reason any change whatsoever would violate God's master plan. As to whether there is any hope of changing the fixed relationships that exist at every level of the natural order, St. Thomas answers with an unequivocal NO. The natural order cannot be made better, nor can individual parts of it be improved. Since the social order, in St. Thomas's cosmological scheme, was regarded as part of the natural order, it only made sense to conclude that the hierarchy of relationships that exist in society was also fixed and unchangeable. Thus, a priest is meant to fulfill his nature as a priest, a lord is meant to fulfill his nature as a lord, and a serf is meant to do likewise. With the emerging mercantile class and the monarchies beginning to challenge the long-accepted pattern of relationships that made up the feudal social order and that preserved Church dominance over European affairs, St. Thomas was no doubt anxious to make the case for perpetuating the status quo.

Lest there be any nagging doubt about the matter of "one's place in nature" and in society, St. Thomas dispels it with one final principle—continuity. According to his cosmology, there are no empty spaces in nature, no vacancies, no holes available to be filled. God the Creator, being totally good and without envy, has filled every single available space in nature's Great Chain. According to St. Thomas, "The lowest member of the higher genus is always found to border upon . . . the higher member of the lower genus."[15] As Albertus Magnus, the great teacher of St. Thomas Aquinas, noted: "Nature does not make [animal] kinds separate without making something intermediate between them; for nature does not pass from extreme to extreme . . ."[16]

If there is no room in nature's hierarchy for maneuvering up or down, since all niches are filled, with each creature performing in "his own manner," it is a certainty that the feudal social order is similarly arranged.

With new political and economic forces tugging at it from all sides, the Papacy was in need of a proper defense of the social order as it had existed for hundreds of years under Church sovereignty. St. Thomas provided that defense. He constructed a concept of nature that was static and in which there was no room for change. Later Voltaire would satirically refer to it as "the best of all possible worlds."

To sum up, St. Thomas's concept of nature proved to be an ideal cosmological companion for the medieval social order. More than anything else, his interpretation of the workings of nature helped legitimize the existing institutional arrangement. For if, as St. Thomas claimed, every living thing in the Great Chain of Being was designed to perform the exact role prescribed to it by the Creator, then the Church fathers and the local monarchs could easily claim that their behavior was merely "fated" and reflected the proper functioning of the natural order as God had set it up. Likewise, any fundamental challenge to the existing institutional framework would necessarily be regarded as an act of indiscretion against the laws of nature.

As people's relationship to their environment has changed, so too have their concepts of nature. Each resulting cosmology has borne the unique imprint of the special circumstances that confronted the human family at a given time and place in history.

All cosmologies, however, share the same overarching theme. They tend to serve as a distant mirror of the day-to-day activity of a civilization. Cosmologies are humanity's way of elevating its behavior to universal importance. It is people's way of convincing themselves that their behavior is appropriate, that it is in accord with the grand operating scheme of the universe. People have always needed to believe that the way they are organizing their life is no different from the way nature organizes itself. Their cosmologies provide a rationale and justification for their every act. Cosmologies provide people with the confidence they need to endure.

Human beings have never felt very comfortable in this world. Their cosmologies have provided them with a thin veneer of security. By means of their cosmologies, people have been able to take an alien and strange environment and make it familiar and hospitable. Cosmologies are humanity's way of saying to itself that it need not worry; that the world it knows, the world it acts in and on, is not very different from the world beyond its grasp. Cosmologies bring the vast universe in line with people's own small corner of it. Without a cosmology people would have no way of knowing that what they are doing is correct. People create cosmologies to sanction their behavior. They have always been quite effective because human beings have never been quite aware of the source of their cosmological formulations. People have always thought that their cosmologies were something handed down to them, or something that they discovered. Never for a moment have they ever believed that they were, in fact, their own creation.

This is not to suggest that people's cosmologies are mere fabrications, as many of the social relativists claim. They would have us believe that our cosmologies have no real footing at all in the external world. The social relativists contend that our ideas about nature are completely subjective and bear no resemblance to the world as it exists in fact. While the social relativists are right in assuming that our ideas about nature are socially biased and deeply influenced by the entire cultural context in which we live, they are wrong in assuming that such ideas are without a basis in the "real" world. The fact is, our cosmologies are based on the workings of the real world, but only that small portion of the real world where society and nature interact. People learn things about nature in the process of organizing it. The things that they learn are useful. They allow people to interact with nature, to manipulate and appropriate it. The problem is that people take the things that they have learned about nature and puff them up in such a way as to create an all-encompassing explanation of the workings of the cosmos. Cosmologies, then, are distortions. They are society's way of inflating its rather limited relationship to the environment into universal truth. Cosmologies are made up of small snippets of physical reality that have been remolded by society into vast cosmic deceptions.

While cosmologies have been an essential feature of human societies from time immemorial, it is necessary to draw a critical distinction between the way pretechnological cultures conceive of nature and the way technological cultures do. In pretechnological cultures people's relationship with nature is still participatory and intimate. Because people's ability to manipulate and redirect nature is slight, their overriding preoccupation is with fitting into the world as it is. In fact, their very survival depends on their ability to conform their behavior to nature's. Their cosmologies reflect their reverential attitude toward the larger forces which control their life. In pretechnological cultures, the cosmologies tend to express the idea of partnership and cooperation between people and nature because that is precisely how people interact on a day-to-day basis with their physical environment.

As human beings begin to develop the mechanical arts, their relationship to nature changes and so too do their cosmologies. With increasing reliance on technology, humanity begins the slow process of wresting itself away from complete dependency on nature. Technology allows humanity to create distance between itself and nature. Technology also gives people power to redirect nature, to channel its forces in new ways. With the mechanical arts humanity can imitate nature, even subsume it. The mechanical arts fundamentally alter people's relationship to their environment. The participatory, intimate union of people and nature is slowly undermined, until it finally ruptures. Increasingly, technology acts as a mediating force between society and the physical world. Humanity uses the mechanical arts to create a "second nature," one cast increasingly in its own image. Its cosmologies, in turn, more and more come to resemble the new technological world it is constructing.

From the first stirrings of the pyrotechnical revolution thousands of years ago, Western cosmologies began to express the emerging technological relationship between people and nature, until today it is fair to say that our cosmologies have become a virtual mirror image of the technological world we have constructed.

In examining the gradual transition in human cosmologies from the time when they represented people's partnership with nature to today, when they represent a mirror image of the technological world we have created, we find an unmistakable psychological pattern developing at the same time, a pattern that becomes more discernible with each successive cosmology and that increasingly influences the very way we come to formulate our view of nature.

This psychological pattern is deeply intertwined with humanity's technology and becomes more pronounced as our technological mastery over nature becomes more sophisticated. What *Homo faber* or "technological man" wants is to overcome death, to be totally self-contained, to disassociate himself from any identification with the rest of nature—and to be able to justify his behavior in the process. These psychological drives exist in a synergetic relationship with technology. As technology becomes more advanced, these psychological drives become more poignant. As these psychological drives become more poignant, they stimulate the drive for greater technological advance. These psychological and technological drives continue to intertwine, and periodically they metamorphose into new cosmological formulations that reflect both humanity's technological activity and its innermost needs and desires.

There are always two natures, then: nature as it is and nature as human beings would like it to be. It is this second nature that is the constant focus of people's attention. Therefore, if we want to understand how our children's generation is going to reconceptualize nature, we first need to examine the basic psychological components that have influenced every one of humanity's cosmological formulations from the time we began our transition from pretechnological to technological beings.

Life, Death, and Immortality

Freud once said: "The goal of all life is death."[17] What he should have added is, "Not if we can help it." Remember that as soon as God expelled Adam and Eve from the Garden, He set up a defense perimeter so they couldn't sneak back in. God was concerned that the pair, already armed with the knowledge of good and evil, might try to appropriate the other important tree in the Garden—the Tree of Everlasting Life. If they succeeded they would become gods themselves.

Every animal shows terror when in danger and will struggle to survive against the threat of imminent death. But only human beings are aware of their own impending demise. In his book *Escape from Evil*, Ernest Becker says that "man wants to persevere as does any animal or primitive organism; he is driven by the same craving to consume, to convert energy, and to enjoy continued experience."[18] The difference, says Becker, is that man "is conscious that his own end is inevitable, that his stomach will die."[19] This awareness places humans in an awkward position vis-à-vis the rest of nature. Our consciousness of our own ultimate death removes us from the immediacy of nature. We can never really relax and flow with the becoming process, like the rest of the animals and plants, because of the perpetual anxiety we carry with us concerning the inevitable outcome of the process.

Our consciousness allows us to look ahead, but we pay the price for the special gift that has been bequeathed to us. Our expanded temporal horizon enables us to predict and control the flow of events in the world, but the better our prophetic vision, the more likely we are to perceive the far-off shadow of our own doom. Anyone who has ever gone to a fortune-teller knows all too well the feeling of wanting to know everything that lies ahead, short of knowing the time, date, and place when he will meet his final fate. If by chance the clairvoyant lets slip exactly when she thinks the grim reaper is going to make his appearance, one suddenly feels cheated, and that everything else in the forecast, regardless of how rosy, was not worth knowing about.

Death has always been unacceptable. It is the only thing in life we can ever really be sure of, but it is also the one thing we hope will never happen to us. This yawning gap between our greatest wish and nature's greatest reality is the foundation on which cultures are erected and cosmologies are modeled. In each case, the cornerstones remain the same.

First we humans build better traps to capture more and more life around us, as if to give us extra breath to ensure our own perpetuity. Much of the human experience is caught up in finding new ways to trap life and appropriate into ourselves the nature that surrounds us. We do this not just to exist but to outlive death. If this were not the case, we would never have advanced beyond the most primitive survival strategies that were certainly adequate to support life. But to live long is not to live forever. Our desire to perpetuate ourselves, in spite of the inevitable, has spawned an insatiable appetite, a lust to incorporate other life into ourselves. Like the ancient trapper, humanity has learned how to lie in wait, to anticipate the movement of life, to predict the comings and goings of seasons and creatures, all in an effort to capture more of the world of nature. Each successive civilization creates larger and larger traps for snaring nature and appropriating it into the human animal. As Becker points out, "Man transcends death . . . by continuing to feed his appetites,"[20] and this driving need finds its way into every one of humanity's cosmologies.

Man the trapper needs to believe in a never-ending stream of life which he can prey on. Therefore, in his cosmologies, nature is always spilling over in plenitude. This was true for the Paleolithic hunter who painted pictures of pregnant animals being speared in the hunt; for Neolithic farmers whose female goddesses were always fecund and fertile; for the Sumerians whose land was bathed in semen and made fertile by the water god, Enki; and for the Greeks and Christians whose created order was satiated with a great multiplicity of creatures filling every possible niche. Human beings have long been obsessed with the idea that nature is robust and inexhaustible, because if it were otherwise, we couldn't be assured of an unlimited supply of life to satisfy our own insatiable needs. The idea of limits and finitude in nature would be unthinkable because it would reawaken humanity's own sense of its personal finiteness and transience.

But human beings can't live by bread alone, so we set out with the morning sun to build castles in the wet sand left by the outgoing tide. We shape and mold with an almost frantic gusto. Never mind that our handiwork will be washed away in the evening tide. We will be long gone from the beach by then, secure in the illusion that we have made our mark, created a world from the world, and left something behind of permanence. Humanity attempts to overcome death by erecting monuments to its own immortality. What we call culture is an exhibit hall full of bric-a-brac and symbols calling out the names of their long-departed inventors. Human beings have created written words that live on in perpetuity. They have erected pyramids to last forever. They have created the notion of corporations and states that enjoy everlasting life. They have even created inheritance to ensure that their past labor will be transferable into the infinite future through their heirs.

Echoing the thoughts of Friedrich Nietzsche, Becker contends: "What people want in any epoch is a way of transcending their physical fate, they want to guarantee some kind of indefinite duration, and culture provides them with the necessary immortality symbols or ideologies; societies can be seen as structures of immortality power."[21]

Human beings need to believe that a world of permanence exists beside, above, or beyond the transient world in which they find themselves. So we build all sorts of monuments to our own immortality as a kind of initiation rite and confidence booster. The more of our handiwork that lives forever, the more convinced we become that we too will live forever. Monuments are our way of continuing to confirm our belief that immortality awaits us in the world to come. Thus, all the most important activities human beings engage in, all the rules and regulations, all the organizing and ordering, all the making and building that comprise what we call culture are activities that somehow help facilitate the rites of passage to the other world of everlasting life.

To a creature always in search of permanence, the earth appears a hostile and forbidding place. Everywhere people look in this world, things appear to be passing by. Life moves to death and growth moves to decay in a world of perpetual movement. That movement is always flowing downstream. Its very irreversibility is a painful reminder that "we were born to die." So we have, through history, attempted to stop the flow downstream by building an array of cosmological dams to hold the becoming process at bay. Every time we succeed in capturing a larger chunk of nature, we attempt to wrap it up with timelessness. We do this by convincing ourselves that the new way we are exploiting nature is in some way linked with the eternal order of things. It is our cosmology that sanctifies our daily manipulation of nature. It convinces us that our behavior is in consort with the workings of the universe. This connection confers an air of permanence on our activity. We come to believe that in our day-to-day activity we are somehow participating in the eternal order in some limited way here on earth. This pretension allows us to conclude that we can continue to do what we are doing forever, without fear of loss, because what we are doing is somehow part of the grand operating scheme of the universe.

Our actions are a little like those of the child who captures a beautiful butterfly and puts it away in a shoebox in the hope of preserving it forever. Every time we capture more of nature, we lock it up in a cosmological shoebox in the hope of preserving it forever.

Self-sufficiency vs. Interdependence

Humanity is single-minded in its determination to endure, to overcome, to perpetuate itself. We seek to live, and we are prepared to pay whatever price is exacted, as long as we can be guaranteed that we will persevere.

To assure our success, we engage in a never-ending war to overcome and capture the becoming process of the natural world. But amid the corpses and the stench of death that we leave behind us on our march to immortality is a voice always reminding us of the toll we exact for a victory that will always remain beyond our grasp. That voice is always there amid the carnage, reminding us of where we too came from, where we too will end up, all the while reproaching us for the suffering and pain we inflicted on a trip that need never have been taken because it always returns to where it began. "From dust thou art, and unto dust shalt thou return."

Humanity is concerned with order vs. disorder and seeks to bring everything under its control. Nature is concerned with unity vs. separation and asks us to surrender to the oneness of which we are a part. Humanity seeks the elation that goes with the drive for mastery over the world. Nature offers us the sublime resignation that goes with an undifferentiated participation in the world around us.

This is the way it has been for Western civilization. Humanity seeks self-sufficiency while nature demands relationship. Humanity wants to be invulnerable, but nature reminds it that it is bonded to all other things. Never is this underlying tension more apparent than when we come face to face with a beggar on the street. On each occasion we are forced to choose between our desire to perpetuate ourselves at all costs and the awareness of our own dependence on all other life for survival. One side of us feels only contempt. The beggar reminds us of everything that is weak and diseased and unfit. We resent his lingering on, and we loathe his feeble attempt to hold on to a life that is already used up. We prefer him dead, so that he will no longer breathe the air meant to fill our nostrils. His death hastens our life. Playwright Eugene Ionesco observes: as long as we are not assured of immortality, we shall never be fulfilled, we shall go on hating each other.[22] Ernest Becker agrees, claiming that "the most general statement we could make is that at the very least each person 'appropriates' the other in some way so as to perpetuate himself."[23]

Yet the other side of us reminds us that this beggar is truly the spoils of our victory. We are indebted to him because our life was gained at his expense. He was appropriated into us so we could live beyond ourselves. We feel remorse and we empathize with him because we know that what he no longer is has been somehow absorbed into us. In this sense we have fused into one another. Part of him is now part of us, and we feel obligated to pay him back for what we appropriated. Psychologists and anthropologists call this dance of the mind "guilt." Guilt can best be summed up as the great human conflict between resentment and a sense of obligation.

We perpetuate ourselves by appropriating others—our fellow humans and other living things. The hard reality is that some things die so that others might live. To quote the Nobel prize-winning author Elias Canetti:

> Nor can we deny that we all eat and that each of us has grown strong on the bodies of innumerable animals. Here each of us is a king in a field of corpses.[24]

For all of us, the question always uppermost on the list of human queries is, How much appropriation is enough? Surveying the history of the human animal and its insatiable quest for immortality, philosopher David Hume remarked, "There is no reason why I should prefer the pricking of my own finger to the death of a hundred human beings." We would much rather that others die than sacrifice ourselves, but we still feel a certain remorse at the taking of another life.

The roots of this dilemma can be traced to the fact that, while we kill, we are the only detached killers in the animal kingdom. We can watch ourselves in the act. We alone of the animals are self-conscious. Other animals participate in the death they exact. Victim and prey fuse as the vanquished is swallowed up into the victor. *Homo sapiens* alone kills at a distance. If distance precludes participation, it offers perspective in return. At a distance human beings can perceive relationships between things over time and space, and because we do, we are caught in a paradox. We prefer to remain detached, knowing all the while that we are bonded to other life. Desiring only to be self-contained, we are, nonetheless, aware of our dependency. We kill to live, but in so doing, we are aware that we are severing our own lifeline. Of the human predicament, ecologist Donald Worster wrote: "If man is in fact bound together in one great organism with all other creatures, then to kill any of them is to commit suicide."[25]

We would like to think of ourselves as self-contained, but in appropriating other life to perpetuate ourselves, we become aware of our utter dependency. The death of another living thing is always a painful reminder of the debt we owe for the next breath we take.

Desacralization

The world-renowned anthropologist Paul Radin once said that "the history of civilization is largely the account of the attempts of man to forget his transformation from an animal into a human being."[26] But we're never totally successful, because we continue to feel a sense of indebtedness. Such feeling would be impossible unless we somehow still "empathized" with the rest of life's creatures. To owe, there needs to be felt a sense of mutuality, of relationship, of interdependence. Yet we resent feeling indebted. It is a painful reminder of our own dependency and vulnerability. So we resolve our dilemma by severing any empathetic association we might feel toward the rest of the living world we are borrowing from. The process is called desacralization, and it has been incorporated into every concept of nature since Western man first broke from plant and animal worship thousands of years ago.

Psychologist Abraham Maslow tells of his first encounter with the desacralization process as a young medical school student. A woman's breast was to be removed. As the electric scalpel began to burn through the flesh, Maslow remembered, "the delicious aroma of grilling steak filled the air."[27] Throughout the operation, the surgeon made carelessly "cool" and casual remarks about the pattern of his cutting . . . finally tossing this object through the air onto the counter where it landed with a plop. It had changed from a sacred object to a discarded lump of fat. . . . This was all handled in a purely technological fashion—emotionless, calm, even with a slight tinge of swagger.[28]

In his book *Where the Wasteland Ends,* historian Theodore Roszak recounts a similar trauma his young daughter experienced when they went into a butcher shop for the first time. She was aghast at the sight of a dead animal dangling from a rope suspended from the ceiling. She had been used to seeing meat cut up in neat rectangular shapes, packaged in cellophane, organized in rows in an open freezer compartment at the supermarket. Since there was nothing to suggest that what she saw in the supermarket was in any way related to animals, she was greatly upset at the discovery that the food she so enjoyed was nothing more than dead flesh taken from an animal carcass.

Humanity cannot afford to acknowledge all of the blood that it spills and the destruction it inflicts on the world in its effort to perpetuate itself. Desacralization is a process that allows us to sever any relationship we might feel to other living things. By draining the aliveness out of things, we can pretend that our control and manipulation are of little consequence. Man the trapper becomes man the taxidermist, disemboweling nature of its spontaneity and movement, and stuffing it with a leaden inanimateness.

People separate nature into two categories—those things outside our control and those things under our control. Those things we can't control we tend to regard as sacred. Philosopher Jacques Ellul says that people pay homage to those things which they cannot overcome.[29] As parts of nature come under our control, we desacralize them, turning them into mere utilities. In other words, the part of the becoming process that cannot be anticipated and manipulated remains in the realm of the sacred; the part that can be anticipated and made to serve human ends becomes profane. Whether a physical phenomenon is regarded as a miracle or not depends on whether or not people are exercising control over it.

It is often observed that we feel awe, fear, reverence, and re-spect before things that we cannot dominate. They are sacred to us precisely because we feel powerless and at their mercy. It is also true that familiarity breeds contempt and indifference. As we gain control over things, they lose their fascination for us. As we incorporate them, they become commonplace. What was once the object of our fear and respect becomes a mundane ap-pendage that we take for granted as we would our own heart-beat. Our respect and reverence for nature diminishes as we gain greater control over it.

The first sight of a wild stallion running across the open tun-dra must have evoked a sense of awe in our Paleolithic ancestors. No doubt many early cosmologies endowed the horse with all kinds of sacred qualities. Today, it's hard to imagine a horse as sacred, now that we have learned how to sit on it, harness it to a plow or carriage, and even turn it into dog meat and glue.

As humanity has brought more and more of nature's suc-cession under its control, it has removed more and more of the things of nature from the sacred column and placed them in the profane column. Our cosmologies read like an accountant's credit-and-debit sheet. Each one itemizes those aspects of nature that we still "owe" our respect and reverence to and those as-pects of nature that "owe" respect and reverence to us. With each succeeding cosmology the sacred column lists fewer entries and the profane column more. Schiller called this process the "disgodding" of nature.

For the Paleolithic hunter-gatherer, plants and animals were imbued with sacred qualities. Not yet having domesticated either, our ancestors humbled themselves before those things in nature that they depended on but that were often outside their influence. With domestication and the transition to agriculture, humanity began the long process of draining nature of sacred qualities.

The Hebrew cosmology marked a significant turning point in the history of humanity's desacralization of nature. In Genesis, Jehovah lets Adam "name" the animals. Humanity put its brand on its fellow creatures, and henceforth they become its charge. Even so, the Hebrew prophets are forever castigating their people for returning to animal worship. British anthropologist A. M. Hocart explains man's struggle to desacralize the animals.

> As his superiority and mastery over the rest of the living world became more and more apparent he seems to have become more and more anxious to disclaim relationship with animals especially when worship became associated with respect . . . man is loath to abase himself before an animal.[30]

According to Jewish cosmology, plants and animals are devoid of souls. The spirit which once permeated nature is now totally removed and exists as a creative force responsible for nature. The sacred quality of this spiritual force resides in one God, above and beyond nature, and is shared, to a limited extent, with only one living creature, the human being, who is made by God in his own image.

The Greeks, in turn, saw all physical phenomena as pale copies of some pre-existent eternal forms. Since everything in the earthly world was only a facsimile of the real thing that existed in the other world of pure ideas, it could justifiably be the object of people's manipulation.

Desacralization is a code word for deadening. Only after a living thing has been thoroughly deadened can it be prepared for assimilation. Unlike the other animals, human beings prefer not to tear into live flesh. We would rather separate the process of killing from the process of eating. We separate ourselves from nature, not only in the pursuit of it, but also in the consumption of it. Regardless of the stage of the assimilation process from pursuit all the way to consumption, we prefer to keep our distance. We adamantly maintain ourselves as subject and reduce everything else to object. Greater separation from nature makes it easier and easier to capture, maim, and kill with impunity. Seeing the vague outline from afar is not the same as hearing the breath rush in and out of the nostrils. In Vietnam B-52 pilots dropped bombs on targets without ever witnessing the consequences. In contrast, our ground soldiers had to hear the tortured screams of the enemy soldiers and of their own comrades.

The bomber pilots have far fewer nightmares today. Detachment and desacralization go together.

To sum up, desacralization is the psychic process humanity employs to drain its prey of aliveness in order to make it palatable. It is our way of convincing ourselves that there is no fundamental likeness between us and other living things. For it is always more difficult to kill and incorporate things that one identifies with. The desacralization process allows human beings to repudiate the intimate relationship and likeness that exist between ourselves and all other things that live.

Morality

If humankind is continually desacralizing the world around it, then where do morality and ethics fit in? After one cosmology after another is appraised, a rather disquieting thought suggests itself: Morality might serve a far different role from the one normally attributed to it. In fact, there's good reason to indict morality as humanity's chief accomplice in the appropriation of nature.

When we speak of morality and ethics we're talking about the rules people are expected to obey in their future actions. We say that a person "should do this" or "should not do that." Shoulds and should nots always refer to future actions. Morality in the form of shoulds and should nots is the community's way of binding its individuals to a prescribed behavior pattern. Morality is how human beings order their own future behavior.

To make sure that everyone is willing to accept the same moral code, it is necessary to claim that its source is beyond human debate and reflects some ultimate truths that govern nature. In every society, then, being good means adhering to the natural order of things, while being evil is associated with resisting or challenging that order. This raises some fairly troubling questions about the role of morality and ethics and the true ends to which they are put. For if doing good means adhering to the natural order of things, and that order, in turn, is just a symbolic representation of the day-to-day activities of the culture, then "doing good" amounts to little more than behaving in a manner that is compatible with the way the society is set up. So for the whole range of behavior that is part of the orthodox routine of the culture, there is never any need to question one's actions, since the cultural norms have been sanctified by having been found to conform with the natural order of things. Questioning comes into play only when individual behavior deviates in some fundamental way from the accepted behavior of society.

In reality, then, ethics are designed to be compatible with the way people organize the world around them. Moral codes keep people's future behavior in line with the way the society goes about organizing and assimilating its environment. Like a sandwich, they reinforce and constrain at the same time. When a moral code says "should," it means that one is expected to behave in a manner that supports the organizing patterns already established by the community; and when a moral code says "should not," it means that one is expected not to deviate in any way from that pattern.

Morality is humanity's way of ordering itself around in that small portion of the world that it has isolated, desacralized, and made its own. Morality transforms what might otherwise descend into a war of each against the other into concerted group activity with a shared sense of purpose. Moral codes serve as a kind of binding contract in which all parties agree to organize the future the same way. When you look at it this way, nothing could be more utterly utilitarian than morality itself. The nineteenth-century English philosopher Herbert Spencer observed that moral codes develop from "the experience of utility organized and consolidated through all past generations."[31]

By themselves, moral codes offer a pretty flimsy basis for ensuring domestic tranquility. It is convenient for everyone to agree on the right way to organize the future, but the hard reality is that the payoff is often slight or negligible for all but that tiny handful of individuals, in every society, whose power and prestige guarantee that most of the future bounty will accrue to them and their heirs. For everyone else, there has to be some other compelling reason to be "good." There is, and it is a reason that influences every cosmology human beings construct. Once again, we come to the idea of plenitude, self-perpetuation, immortality. "Do good, and you will be rewarded. If not in this world, then in the next." Echoing Nietzsche, Ernest Becker once remarked:

> All morality is fundamentally a matter of power; the power of the organism to continue existing by reaching for superhuman purity . . . the earning of spiritual points is the initial impetus of the search for purity, however much some few noble souls might transmute that in an unselfish direction.[32]

By being good we ensure ourselves a place in the world beyond. We rack up points, we build up immunity, we store up power. We give ourselves up, body and soul, to the task of organizing the immediate future in return for a reward in the far-distant future. That reward is eternal life. Ernest Becker sums up humanity's need for morality:

> Purity, goodness, rightness—these are ways of keeping power intact so as to cheat death; the striving for perfection is a way of qualifying for extraspecial immunity not only in this world but in others to come.[33]

Good, in every culture, is associated with eternity. Since everlasting life is our ultimate wish, it only makes sense that we would come to associate it with the ultimate good. We aspire to be good because we aspire to become part of eternity. Evil, being the opposite of good, is associated with transience, with earthiness, with disease, decay, and death: all of the things we hope to overcome. In every society the battle between good and evil is the battle between immortality and mortality. Goodness is humanity's ultimate weapon in its struggle to overcome nature and the limits imposed by it. The finiteness of nature is a constant unpleasant reminder of our own precarious existence. It's no wonder then that for much of human history we have viewed nature with a sense of utter dread and have come to see in nature all sorts of evil forces threatening to do us harm. Nature in its very transience, in its finiteness, becomes something to overcome, to defeat. It becomes the enemy. As a

result, the battle between good and evil and the battle between man and nature have come to mean the same thing.
Taken to its extreme, then, the ultimate triumph of good over
evil becomes the complete and total vanquishment of this
earthly world. To purify and prepare ourselves for the infinite,
we must, of necessity, expunge all vestiges of the finite. So we
tear into everything around us, devouring our fellow creatures
and the earth's treasures, all in the name of doing good, of
ridding the world of evil. What we are really ridding the world
of is its aliveness. Goodness becomes a mask for our nihilism. It
is as if we were determined to destroy every last reminder of
this finite world in the hope of ridding ourselves of the painful
awareness of our own temporary nature. Morality is humanity's
chariot to heaven. It leaves behind it a dust storm of destruction.

One more thing need be said at this juncture about concepts of
nature, perhaps the most disturbing thing of all. Our cosmologies are not so much a representation of nature as they are a re-
pudiation of it. This has been so ever since Western man and
woman first began to use the "mechanical arts" to reshape the
world in their own image. Every cosmology is touted as an explanation of the workings of nature. In reality, every cosmology
is an expression of humanity's overwhelming desire to negate
everything that is alive and finite. That's because the way we
come to view nature is heavily influenced by our psychological
needs, and these needs conspire against nature. We seek immortality while nature is mortal. We drive for self-containment
while nature is made up of relationships. We prefer distance
while nature is participatory. Concepts of nature do not so much
reveal nature as they do our compulsive desire to escape from its
clutches.

It could not be otherwise. We know nature by the way we interact with it, and we interact with it in order to appropriate it.
Therefore cosmologies tell us more about how people are
"using" nature than they tell us about nature itself. Because our
relationship with nature continues to be an extractive one, our

cosmologies will continue to portray nature in terms that are congenial with the way we go about devouring it.

Although they are always proclaimed at the time as invincible, every cosmology has eventually been either abandoned altogether or greatly modified to be compatible with changing social conditions. That's because concepts of nature are so intimately tied to the day-to-day activity of a culture. But we know that when qualitative or even catastrophic changes in the environment occur, they often overwhelm the existing pattern of relationships between people and their environment. When this happens, radically new ways of structuring life take hold. Suddenly the old concept of nature becomes the object of criticism and revision. Eventually it comes to look quite ridiculous, and by the time a new concept, more in line with the new circumstances, begins to take hold, the old concept becomes the butt of titters and jokes and finally is relegated to the ultimate disgrace—it's all but forgotten.

Darwin's theory of evolution has, for the most part, proved to be a very compatible companion to the Industrial Age. It's hardly possible that it could have survived and prospered as it has—even in its modified form—if its tenets had been grossly at odds with the basic operating assumptions of industrialism. But what now? The Industrial Age is entering its mature, if not senescent, stage. We are beginning to make the transition from a resource environment based on nonrenewable energy to one based on renewable resources. This revolutionary shift in environments is already fostering the incipient rise of new technologies, new institutions, and a realignment in social relationships. We are moving out of the age of pyrotechnology and into the age of biotechnology. For the first time in history, people are about to engineer the biological design of life itself. It is no accident, then, that just as humanity begins to cross the divide into a new epoch, the debate over Darwin's theory of evolution is beginning to smolder and threatens to ignite into a firestorm of controversy, recrimination, and reassessment.

Today, Darwin's theory is coming under increasing attack from inside and outside the scientific community. Hardly a week passes without another scholarly journal or popular magazine entering the fray with its own attempt to show where Darwin went wrong. While the scientists and social commentators are busy focusing their attention on the rather narrow debate that is emerging around Darwin's theory of evolution, they are missing the real significance of the battle that's beginning to brew. To question Darwinism is to question much more than a theory about the origin and development of species. What's really at stake in the current debate over Darwin's theory is a way of life, a way of relating to the world.

Even before we are old enough to uphold an opinion, we are weighed down with a catechism laden with Darwinian aphorisms. We are informed that the world out there is a jungle, menacing and hostile. We are taught that it's a "dog eat dog" world and that our very existence depends on our being more fit than our peers in the struggle for survival. We are made to believe that a modicum of greed is natural and that by advancing our own self-interest we are adding to the common good. We are warned that change is inevitable, but that it should come in small, tempered doses, not suddenly and unexpectedly. We are told that while chance enters into all things, success comes to those who are able to take advantage of opportunities as they arise. If there be any temptation whatsoever to challenge these basic Darwinian tenets, we are quickly admonished with the all but conclusive dictum that, try as we will, we can't fight "human nature."

So accustomed have we become to thinking of the world in a Darwinian mode that it is painful to even consider the possibility that the Darwinian conception might be in error. That is why attacks on Darwin's theory, regardless of which quarter they come from, are generally greeted with incredulity, if not outright hostility. Yet there is no doubt that such attacks are going to increase in the years ahead, and eventually they will triumph, leaving Darwin a lifeless corpse, a distant memory of a bygone era. To understand why this is the case, we need to do a bit of detective work, what scholars refer to as social reconstruction. We need to discover exactly how Darwin came up with his theory in the first place. What we are going to find is that his inspiration was not unlike the inspiration that influenced the other great thinkers of his age. What Darwin discovered was not so much the truths of nature as the operating assumptions of the industrial order, which he then proceeded to project onto nature. Is it any wonder, then, that the passing of the industrial era should also mark the passing of Darwin's concept of nature? Our way of life is about to be "changed, utterly changed." So too is our way of looking at the world. Before exploring what lies ahead, however, we need to understand better what we are leaving behind.

PART THREE

DARWIN'S VISION

A Reflection of the Industrial State of Mind

It is no secret that Darwin's theory of evolution has been exploited over and over again to justify various political and economic ideologies and interests. Social Darwinism has been examined, debated, and analyzed for over a hundred years. In virtually all the discussions of Social Darwinism there is an underlying assumption that the theory itself is a disinterested, objective, impartial recording of nature's operating design, untainted by social context and cultural bias. It is assumed that what Darwin discovered is a law of nature and that society then exploited it for political ends. A new generation of scholars, however, is beginning to question the theory itself, suggesting that in its very conception it might have been as socially biased as the ends to which it was later used.

Otto Rank suggests that Darwin's theory was just the English bourgeoisie looking into the mirror of nature and seeing his own behavior reflected there.[1] Although such pithy comments are unlikely to grace the pages of any introductory book on biology, it remains a fact that Mr. Darwin was indeed a product of his time and subject to the flights and fancies that embroidered the Victorian landscape. It can hardly be a matter of doubt, says

University of Connecticut historian John C. Greene, that, "like
every other scientist, Darwin approached nature, human nature,
and society with ideas derived from his culture . . ."[2] That being
the case, to understand Darwin's theory of biological evolution it
is necessary to understand the economic, social, and political environment that provided the imagery that he used so artfully to
sketch his "creation."

Darwin's life spanned the very years that marked the transition from an agrarian economy to the Industrial Age of capitalism. England was at the forefront of the revolutionary changes
that were transforming the economic life of Europe. Having a
head start on her Continental neighbors, she needed a new cosmology that could make sense out of and be compatible with the
disorienting array of economic changes that were turning merry
old England from a land of haystacks into a land of smokestacks.
John Greene jokingly asks why "nearly all of the men who pro-
pounded some idea of natural selection in the first half of the
nineteenth century were British."[3] "Given the international
character of science," Greene says, he finds it more than a bit
". . . strange that nature should divulge one of her profoundest
secrets only to inhabitants of Great Britain."[4] Of course Greene
isn't really surprised at all. He would be apt to agree with the
sentiments of biologist Alexander Sandow. Writing in the *Quarterly Review of Biology,* Sandow observes that "Darwinism sprung
up where and when capitalism was most strongly established."[5]
While such an observation is likely to ruffle the feathers of most
self-respecting scientists, for many scholars of history it is difficult to imagine that the appearance and acceptance of evolutionary theory and the rise of industrial capitalism are merely
coincidental in time and space. The historian is likely to concur
with Greene that "British political economy, based on the idea
of the survival of the fittest in the marketplace, and the British
competitive ethos generally predisposed Britons to think in
terms of competitive struggle in theorizing about plants and animals as well as man."[6]

The economic changes taking place in England and on the
Continent in the first half of the nineteenth century were more
uprooting and far-reaching than any experienced by humanity

since the advent of full-scale agriculture and the formation of the first city-states at Sumer. After thousands of years of agrarian existence, the human family was embarking on a new course. Millions of people abandoned their plows in the very fields their ancestors had tended from time immemorial. They gathered their meager belongings and marched right out of the sunlight into darkened factories, where they fastened themselves to a new tool, the industrial machine. After working for millennia with their backs to the sun in the open fields, waiting patiently for it to bathe their land with its magical energy, they now leaned forward, shoveling stored sun into giant blast furnaces, their faces aglow from the radiance being released from the tiny packets of energy capital. With coal, people no longer had to wait. Their sun now came from deep beneath the ground. For the first time in history, the sun waited on people, to extract, to transform, to harness.

The transition from an agrarian existence to industrial capitalism was as much the transformation of our energy base from solar flow to solar stock. Up until this time, human existence had been conditioned by the rising and setting sun. Biologically designed to move in concert with nature's own cycles, people could never escape much beyond the bounds dictated by the sun's constant and never-changing rhythm. With the discovery and harnessing of stored sun, humankind broke through the barriers. People now had a batch of stored fire at their disposal, and with it they could heat up the bowels of the earth, turning the planet's material into a cornucopia of economic utilities. By dipping into and utilizing a stored-up supply of hundreds of millions of years of the sun's energy, humankind greatly accelerated the pace of economic activity.

The sun is the source of all animation. It is the energy catalyst that brings matter to life. It is the agent responsible for mutation and change. The sun's flow is virtually constant. Its regularity assures a generally steady movement within nature itself. The use of stored sun, however, effects consequences of a very different kind in nature. It is concentrated and it is utilized by people selectively. It can speed up activity within an isolated area way beyond what would be possible by the steady bath of the sun's flow alone. Stored sun, then, can be used to greatly accelerate movement of all kinds. When it was transformed by use of a new invention, the steam engine, that's exactly what happened. The Industrial Age is characterized by a profound acceleration in movement and a quickening of pace in every aspect of life. This acceleration of life brought with it a steady barrage of novelty that required far greater foresight and planning than anything humanity had ever dealt with before. As people stretched their conscious minds to absorb this new reality, they also stretched their concepts of nature to accommodate the expanded temporal horizon they were now forced to operate in. It is this basic alteration of time and space that conditioned the new economic environment and that led to the gradual disintegration of St. Thomas Aquinas's Great Chain of Being and its replacement with the theory of evolution put forward by Darwin and his contemporaries. As Cambridge scholars Mikulas Teich and Robert Young point out, "The first [cosmology] was suitable for a relatively static and rural economy while the other was developed for a rapidly changing and industrializing society."[7] Before examining the interior aspects of that change of cosmology, it might prove useful to look at the many surface changes in English economic life that inspired Darwin's machinations.

Darwin was born in Shrewsbury, England, on Feburary 12, 1809. He first outlined his theory of evolution by natural selection in 1842. During the intervening years Darwin witnessed the wholesale transition of English economic life. The times were turbulent. For the average Englishman it must have appeared that the whole world had gone mad. A new economic order was forcing itself out of the ancient agrarian womb, and the labor pains were intense. Building the industrial infrastructure took large amounts of capital and cheap labor. A few made fortunes midwifing the birth of the machine age. For most, these decades were punctuated with sacrifice and suffering on a monumental scale. While the masses gave their backs over to building the factories, machines, and railroads, it was the rising bourgeois class of entrepreneurs in alliance with the older, moneyed class of English gentry who profited. A character in Disraeli's *Sybil* described Victoria's England as "two nations; between whom there is no intercourse and no sympathy . . . the rich and the poor."[8]

According to English historian Richard D. Altick of Ohio State University: "The gulf was widening year by year, and the search for the means of narrowing it constituted the great challenge of the time."[9] Indeed, there was simply no precedent for what was taking place, no way to appeal to tradition, for what was happening was utterly alien to humanity's past experience. Millions of men, women, and children were made homeless overnight. They were refugees of an economic order that had suddenly collapsed after 1,500 years. Forced to make their way from a country environment of thatched huts and small villages that still bore a striking resemblance to their Neolithic origins, they found themselves suddenly surrounded by tar and concrete, row after row of makeshift shacks, belching chimneys spewing black smoke over the heavens, the deafening sound of pulleys and pistons and wheels. It was a world that was worlds apart from anything human eyes had ever fixed on.

The masses came streaming in from the country to build and run the giant machines of the capitalists and to suffer humiliation and bondage under the relentless drive for increased production.

> Faster and faster, relentlessly . . . blindly . . . the Industrial Revolution put an end to eighteenth century rural order . . . workers massed like swarming bees round those areas where iron and coal and other minerals lay beneath the surface. . . .[10]

These years saw the rise of the modern city. Like giant magnets the new cities of iron and steel attracted the scattered trailings of rural farmers, yeomen, and laborers, heaping them up one on top of the other in numbers that defied history's imagination. According to Richard Altick: "The population of all the major industrial conurbations, Manchester, Leeds, Bradford, Birmingham, Liverpool, and Sheffield, gained an average of 50 per cent in the single decade 1821–1831."[11] Another noted historian, A. F. Weber, concludes that the concentration of people in cities was "the most remarkable social phenomenon of the century."[12]

Nobody planned the Industrial Age. It just happened, and with such lightning speed that no one knew exactly how to respond. As a result, says philosopher and historian Edward Manier, "perhaps no period in British history has been as tense"[13] as these years stretching from 1830 to the late 1840s. According to Manier, "Real income per capita fell for the first time since 1700; a decline in the quality of their diet led to a mood of hopelessness and hunger among the poor."[14]

It's hard to know which was worse, the new working conditions of the factories or the living conditions in the makeshift city slums—or, more appropriately, the sleeping conditions—for most people spent most of their waking hours tending the machines. Men, women, and children were forced to work sixteen hours a day and a six-day work week under conditions that ranged from disgraceful to appalling.

> They were deafened by the noise of the steam engines and the clattering machinery and stifled in air that not only was laden with dust but, in the absence of ventilation, was heated to as high as eighty-five degrees. The workers were driven to maximum output by strict overseers, fined for spoiling goods, dozing off, looking out the window, and other derelictions, and forever imperiled by unguarded shafts, belts, and flywheels. Industrial diseases and those caused simply by the proximity of many unwashed, chronically ill human bodies conspired with accidents to disable and kill them.[15]

The factory whistle spelled little in the way of relief for these early victims of the Industrial Age. Weakened to the point of utter physical exhaustion, workers retreated to dark, windowless hovels, where it was not uncommon for seven or eight people to share a common bed. Often they lacked even the most rudimentary forms of sanitation; garbage and human excrement lay in open gutters, seeping into the living quarters. The stench was often overwhelming. The pollution from industrial waste and human garbage was so thick that "land birds could ride on the surface."[16] The only thing that seemed to flourish amid the squalor was disease. According to health records, in one year alone 16,437 people died of typhoid and cholera. The very young suffered the worst. "One out of every two babies born in the towns died before the age of five."[17]

The effluence of city slums was in sharp contrast to the rising affluence of the new entrepreneurial class. This was a time for fortunes to be made. New names were added to the official register of important personages as the struggling bourgeoisie staked out its claim for recognition and status in British society.

The new, "self-made man" differed in one very important respect from the landed gentry that had for so long dominated the affairs of the nation. The latter group had lived off the inheritances of their ancestors, continuing to reap financial benefits from bounty first exacted by their more scurrilous forebears. They were a rentier class, steeped in the tradition of the old manor estates. Bound by tradition and reinforced by the military power of the monarchy, they were able to levy taxes in the form of rent from generation to generation, with the peasants asking few questions and mounting even fewer challenges to their power. Their wealth flowed from the land, from agriculture, and from the hard labor of the peasant farmers, who gave over much of their surplus in return for the "privilege" of renting soil to raise their crops and pasture their herds.

The new bourgeois class of entrepreneurs also made their wealth off the masses of people. But their work force was the industrial proletariat. Instead of growing food, they were making durable goods. The industrial entrepreneur was surrounded by merchants, traders, and professionals. Together, they created new markets and new services.

As is so often the case in history, the traditional ruling order found itself confronted by a new challenger for power. The result, says Altick, was that

> on the one hand there was a closely interlocked aristocracy and gentry, living as they had done for centuries on the proceeds of their landed estates; on the other, a rising middle class and a more populous working class, both coming to be based more and more in new-style industry and commerce. The landowners wished,

naturally, to cling to their inherited position and prerogatives; equally understandably, the middle class and the workers demanded a voice in political affairs commensurate with the ever larger contribution they were making to the nation's wealth.[18]

The cry for reform echoed from left to right, only to be met with the call for tradition from the other direction. The voices became more shrill in the 1840s as a devastating economic depression heightened the political frenzy on all sides. There was talk of bloody revolution in the streets. People were organizing and arming themselves, factions were jockeying for position, forging new political alliances, and then suddenly, unexpectedly, it all stopped. As one social commentator put it, "Politics went into hibernation."[19] England had weathered its most turbulent political storm in history and now found itself gliding through peaceful waters. The turnaround, however, was in no way attributable to superior statecraft. It was not so much that wiser heads had prevailed as it was that capitalism had finally been weaned. After decades of nursing, industrial capitalism was ready to feed itself. If its appetite was insatiable, its generosity was at least adequate to accommodate the stretched-out hands and open mouths of the English citizens. Prosperity was right around the corner.

The 1850s and 1860s were boom decades. British exports grew more rapidly in the first seven years of the 1850s than in any other period in the nation's history.[20] Employment grew; so did wages, although not as fast as profits. Still, the "trickle down" was enough to mollify the working class, at least for the time being. Britain became the symbol of the Industrial Age and the envy of every nation in Europe. And it had achieved its new status without civil war or violent revolution. Quite simply, the working people were bought off by the new affluence, leaving the poor without allies, to fend for themselves. The entrepreneurial class, in turn, traded away much of its political ambition in exchange for acceptance by the traditional aristocracy it had previously challenged. This, perhaps, is the most curious part of the English economic transformation and deserves a few words.

While the new bourgeois class often chided the landed gentry for its nonproductive posture, it was anxious, nonetheless, to emulate much of the genteel finery that went along with the trappings of established wealth. The gentry, for their part, were not beneath making a fast killing in the bourgeois marketplace, and as historian Martin J. Wiener points out, they began to draw "an increasing proportion of their incomes from railways, canals, mines and urban property."[21] So the bourgeois class came to emulate the cultural style of the gentry, and the gentry, in turn, began to emulate the economic practices of the bourgeoisie. The political result, says Wiener, was accommodation rather than social upheaval.

Charles Darwin

Sitting on the sidelines, watching this historical spectacle unfold for six long decades, was Charles Darwin. An observer by nature, Darwin watched, took notes, and pondered. Like his contemporaries, he was anxious to make sense out of this new world being born.

Everywhere he looked, things were in flux. Englishmen were changing their environment and themselves with deliberate speed. Who would have believed just a generation earlier that messages would be sent over wires, moving hundreds of miles in a matter of seconds. If the telegraph was a source of wonderment, the railroad was even more astonishing. Imagine a machine running down an iron track on its own steam at speeds never before experienced by human beings. By all accounts, an Englishman might well suspect that, for the first time in history, humanity was truly conquering time and space, bringing the forces of nature under control once and for all.

Darwin was undoubtedly impressed. British ingenuity had been translated into British technology, and together they were creating the world anew, if not in six days, then at least in six decades. To the question What was England? Darwin would likely agree with Samuel Smiles's description of the island as a giant machine hissing, belching, and churning out a new, more durable, mechanical world. It was the machine age. Everywhere there was talk of mechanical inventions, and everywhere mechanical inventions were remaking the world. The machine was the single most absorbing preoccupation of the period. In 1851, the first World's Fair was held in London. It was called the Great Exhibition of the Works of Industry of All Nations, and mechanical inventions of every kind were housed in the giant Crystal Palace, made of iron and glass. The whole world was invited to look in and admire the symbols of a new age.

If the machine represented the physical embodiment of the age, the struggle for survival and material gain represented the spirit. Everyone was fighting for a piece of the new order. Beneath the literary image of the proper English gentleman and the simple, respectful English worker was a tempest of angers and hostilities, lurching forward, flailing in every direction. Englishmen were jostling Englishmen, clawing at and trampling over one another in an effort to keep up with the quickening pace of the machine, each intent on assuring that a portion of its output would be his for the taking. The strides and temperament of the period were not lost on Darwin. On the contrary. What made Darwin's cosmology so terribly engaging was that it fit the age as effortlessly as a well-greased ball bearing fits a joint. Says biographer Geoffrey West:

> In the machine age he established a mechanical conception of organic life. He paralleled the human struggle with a natural struggle. In an acquisitive hereditary society he stated acquisition and inheritance as the primary means of survival.[22]

After examining Darwin's many notebooks, journals, and formal publications, any disinterested observer comes to an unmistakable conclusion: Darwin dressed up nature with an English personality, ascribed to nature English motivations and drives, and even provided nature with the English marketplace and the English form of government. Like others who preceded him in history, Darwin borrowed from the popular culture the appropriate metaphors and then transposed them to nature, projecting a new cosmology that was remarkably similar in detail to the day-to-day life he was accustomed to. He was congratulated for unlocking the secrets of nature when, in point of fact, it would have been more appropriate to praise him for having satirized the English frame of mind as it made its way into the Industrial Age. A look at the man and the theory tells the story of this most recent addition to the gallery of cosmological portraits.

Very bourgeois! That's how historians generally categorize Charles Darwin. Speaking of his pedigree, West notes that both sides of the family were

> profoundly middle-class, rising by prudence and steady worth from yeoman to gentlemanly rank, their typical members sturdy individualists, men of serious, even solemn, turn of mind, independent in opinion, not disregardful of cultural and spiritual matters, but intent on making their ways in the world against whatever odds.[23]

By his own accounts, Darwin remembers his childhood as a pleasant though undistinguished time. Quite simply, he was a rather ordinary English lad of middle-class background, and there was little to suggest that he might one day turn the world upside down. Certainly he showed no special talent as far as intellect was concerned. Quite the contrary, he was a modest—if not slow—learner; and his father worried for a time that he might not possess the mental acumen necessary to pursue a professional career. As a youngster, Charles showed "little initiative or ability to grasp principles,"[24] and as his many biographers suggest, "He absorbed the values of his environment as he absorbed the basic ideas of the *Zoonomia,* unconsciously."[25] In his autobiography, Darwin recalls two things about himself as a child that were worthy of mention. First, he enjoyed the role of prankster.

> I may here also confess that as a little boy I was much given to inventing deliberate falsehoods, and this was always done for the sake of causing excitement. . . . [For example], I told another little

boy . . . that I could produce variously coloured Polyanthuses and Primroses by watering them with certain coloured fluids, which of course was a monstrous fable, and had never been tried by me.[26]

Darwin says that his other great enjoyment was "collecting" and that at an early age this propensity was already well developed. Of this particularly absorbing interest Darwin later wrote:

The passion for collecting, which leads a man to be a systematic naturalist, a virtuoso or a miser, was very strong in me, and was clearly innate, as none of my sisters or brothers ever had this taste.[27]

After a public school education, residence at Cambridge, and the famous sea voyage on the *Beagle,* Darwin married and settled down to a rather reclusive living arrangement in the little rural village of Downe. There, safely nestled among forty or so large cottages and a tiny country church, Darwin raised his family. Darwin said he "had never been in a more perfectly quiet country."[28] It was here, far removed from the bustle and clatter of London, and surrounded by the calm serenity of country gardens, that Darwin penned his version of how the world began. It was one of the paradoxes of Darwin's life, notes Geoffrey West, "that while he lived so generally secluded a life, isolated always at least from the industrial struggle typical of the time, he yet stated a theory not to be detached from and singularly expressing the spirit of the century."[29]

But this wasn't the only paradox. Though Darwin's theory of the origin and development of species centered on the survival of the fittest, he was plagued by chronic ill health, so severe that it constrained every aspect of his daily existence. According to historian Gertrude Himmelfarb of the City University of New York:

Suffering was the motif of Darwin's life, as surely as science was its motive. It was, as his son later recalled, "a principal feature of his life, that for nearly forty years he never knew one day of the health of ordinary men, and that thus his life was one long struggle against the weariness and strain of sickness."[30]

The irony of his own personal situation was not lost on Darwin. In musing over his perpetual ill health and frail condition, Darwin once remarked, "It has been a bitter mortification for me to digest the conclusion that the 'race is for the strong,' and that I shall probably do little more, but be content to admire the strides others make in science."[31] Nausea, stomach disorders, dizziness, heart palpitations, and general weariness were so much a part of Darwin's life that they became his normal disposition. "Many of my friends, I believe, think me a hypochondriac."[32] Whether they did or not, and whether it was the case, is a subject that has fascinated biographers and psychologists alike. Of Darwin's mental state and physical disposition, little more is likely to be learned with the passage of time. But certainly it is fair to say, as Darwin himself did, that his ever-present ill health affected his outlook and, to some extent, his imagination.

Equally ironically, the man who popularized the idea of a struggle for survival in nature was himself spared from having to engage in combat to secure his own. Darwin never earned a living. Instead, he lived modestly well throughout his entire adult life on the inheritance left to him by his father. According to Himmelfarb, Darwin felt "a sense of guilt at not earning his keep, and an acute sense of inadequacy at not being able to do so."[33] In fact, the whole idea of making a living seemed "so formidable, so unthinkable, that he could not imagine anyone dear to him capable of it, least of all a Darwin who must have inherited his own constitutional weakness."[34] Evidently his brother Erasmus agreed. He once remarked that "the chances are against any of our unfortunate family being fit for continuous work."[35]

This, then, is the man who gave expression to an age, the man who provided a vast cosmological blueprint to guide (and rationalize) the affairs of industrial civilization. A man isolated from the battle, in ill health, and without the means of securing his own physical survival. Perhaps it could be argued that Darwin was "best fit" to advance the idea of natural selection and survival of the fittest in nature precisely because of his own shortcomings. Indeed, this would make sense if one views human cosmologies as expressing people's innermost hopes, expecta-

tions, desires, and needs. Darwin expressed the convictions of his
time, and as we shall see, those convictions had far less to do
with nature's workings and far more to do with the way society
went about its own work.

Artificial Breeding and
Natural Selection

Talk of evolution had been in the air for several decades. Darwin
brought it down to earth and transformed it from idle banter to
acceptable doctrine by the use of a few very convincing metaphors.

Darwin looked to the way people were organizing nature for
clues as to how nature itself might operate. The revolutionary
changes in agricultural practices provided a starting point for
his observations. In the 1750s, the British began to radically
overhaul their agricultural system. Under pressure to feed an increasing population, a host of structural and technological reforms were pursued. The old open-field system of agriculture
that had prevailed from the early Middle Ages was overturned.
Parliament enacted a series of private enclosure acts, and in the
eighteenth and nineteenth centuries "about six and a half million acres of English landscape were transformed into a ration-
ally planned checkerboard of squarish fields enclosed by
hedgerows of hawthorn and ash."[36] The agricultural reformers
marshaled their enthusiasm under the popular banner "Make
two blades of grass grow where one grew before."[37] The great
strides in agriculture were matched by new developments in the
art of artificial breeding, and it is here that Darwin turned his
attention.

Before 1750, there was little, if anything, in the way of definite
breeds among British livestock. Then, in the 1760s, Robert
Bakewell "began the systematic improvement of sheep, oxen,
and horses and soon produced the famous breeds of Leicester-
shire sheep and Dishley cattle."[38] As the art of breeding became
more sophisticated, it spread from animals to plants and began

to play a central role in the new commercial agriculture system. By artificial breeding, British farmers were able to greatly improve the quality and output of their domesticated animals and plants, and of course, better quality translated into larger profits.

Darwin was absolutely captivated by the success of the breeders. Convinced that their practices held an important clue to the workings of nature itself, he set upon the task of collecting facts about every aspect of the art. Darwin later acknowledged his intuition that by "collecting all facts which bore in any way on the variation of animals and plants under domestication . . . some light might perhaps be thrown on the whole subject [of the origin of species]."[39] In 1837 he became a member of several breeders' associations. Over the next fifteen months he watched, conducted interviews, took copious notes, all in an effort to better understand the process of artificial selection. Two things caught his attention. First, artificial breeding resulted in great variation within a species. Second, certain of the varieties proved more "useful" than others and were therefore preserved in the form of special breeds. Those less useful varieties were simply not bred, and so their traits were not passed along to offspring. According to Alexander Sandow, Darwin reasoned that if "species-types could be changed by people through artificial selection, and the laws of these changes could be discovered, it was but a step . . . to apply these findings to the process of evolution in nature—past as well as present."[40]

Still, there were two giant hurdles to overcome if Darwin was to successfully make the leap from artificial to natural selection. In regard to domestic breeding, Darwin had to concede that while it was possible to create an infinite number of varieties from a given species, no one had ever successfully produced a brand-new species by way of artificial selection. In other words, while it was certainly a proven fact that one could breed hundreds of varieties of cows, no one had ever mated two cows into something other than another cow. While he lamented that "man's power in making breeds is limited,"[41] he provided himself with a glimmer of hope by suggesting that such might not always be the case. Here is the critical point where Darwin strayed from what could be proved by observation and the naked eye to what had to be imagined and conjectured by a fertile mind. Given enough time—a great deal more time than human beings' own limited historical observation could provide—was it not possible, he mused, that the slow process of variation heaped upon variation could result in one species turning into a wholly new and unrelated species? Of course anything is possible, and so Darwin could at least take comfort in the notion that his speculation about the transformation of species, while not provable, was not unprovable either.

Equally troubling for Darwin was the question of how natural selection might take place. As far as artificial breeding was concerned, there was never any doubt as to motive or mechanism. As Darwin himself put it, "Hard cash paid down, over and over again, is an excellent test of inherited superiority."[42] Breeding was designed to select those varieties that would prove most useful in the marketplace. The goal was better milk, wool, meat, and cereal. Yield and output conditioned the entire process. Breeders selected those traits that would be most "useful" in enlarging their profit margins. In other words, "adaptive" traits were those traits that would assure the biggest return on investment. But as Michael Mulkay points out, "It does not follow necessarily from this that variation in natural settings is also adaptive."[43] After all, one would be hard pressed to assume that the motives undergirding human acquisitiveness find their counterpart in nature. Yet it is characteristic of humanity's anthropocentrism that it should always attempt to rationalize its own behavior by projecting it onto the rest of nature, and in this regard, Darwin was no better or worse than the long list of cosmologists who preceded him. Darwin assumed a parallel of sorts to exist between artificial and natural selection.

To begin with, Darwin assumed that just as breeders selected for new traits on the basis of "usefulness," a similar process must somehow be at work in nature. Darwin had no trouble convincing himself that plants and animals in a state of nature must be "selected in accordance with external requirements,"[44] just as domestic livestock are. Of course, this was speculation, but, as far as Darwin was concerned, a very powerful speculation. After all, everywhere he looked he saw the proliferation of ever more useful new varieties of livestock and plants. "Can we wonder, then, that Nature's productions should be far 'truer' in character than man's productions; that they should be infinitely better adapted to the most complex conditions of life, and should plainly bear the stamp of far higher workmanship?"[45] Darwin was absolutely sure that the way people were organizing nature must somehow reflect the process by which nature organized itself.

But then the final question to resolve was, What was the mechanism used in nature to select, and what traits were selected for? Darwin found the answer just fifteen months after he began his work on artificial breeding. In his autobiography, Darwin writes that in October 1838 he "happened to read for amusement 'Malthus on Population' . . . and it at once struck me that . . . I had at last got a theory by which to work."[46]

The Reverend Thomas Malthus, who had published *An Essay on the Principle of Population* in 1798, was greatly "disturbed by the severe famines of that period, by the social displacements due to rapid industrialization, and by the swelling numbers of paupers requiring relief from the local rate-payers."[47] These concerns motivated his famous law on the relationship of human population to resources. Malthus argued that while food production can increase only by arithmetic progression, population grows at a geometric rate. As a result, population always ends up overrunning the available food supply. The constant pressure of population on the available resource base is kept in check "through the elimination of the 'poor and inept' by the ruthless agencies of hunger and poverty, vice and crime, pestilence and famine, revolution and war."[48] In other words, in the struggle for survival, nature ensures that the strong will triumph and the weak will perish.

Malthus is the gentleman responsible for branding economics "the dismal science." That is not to say that he did not offer a morsel of hope. While Malthus was quick to assert that the human appetite would always outstrip its available diet, he did contend that this was a very natural part of God's master plan. The constant struggle between superior reproductive capacity and modest resource supply provided the all-important edge so necessary to sharpen humanity's ingenuity and prowess. As Donald Worster points out, Malthus was convinced that

> without such a harsh decree, man would long ago have relaxed into the sloth of savagery. Only by the threat of hunger has he been stimulated to exert his full capacities and to advance toward civilization.[49]

To Malthus, nature was like a very dependable straining process. It assured that the strong, the productive, the industrious would prevail and that the weak, the lazy, and the slothful would perish. Malthus gave concrete mathematical expression to the popular opinion of the day. In salons and countinghouses the bourgeois class was carrying on as if convinced that in nature, as in society, "survival of the fittest" was the pervasive rule of thumb. Opposed to government interference in the economy, the rising industrial class was cocksure that if everything was just left alone to work itself out in the open marketplace nature would assure that the most industrious would prevail and that the undeserving would not. Malthus put forth what he regarded as an ironclad law that provided the necessary proof that the bourgeois class was looking for to bolster its confidence in the innate superiority of laissez-faire economics.

Darwin was quick to latch on to Malthus's insight. In fact, as Bertrand Russell later pointed out, "Darwin's theory was essentially an extension to the animal and vegetable world of laissez-faire economics, and was suggested by Malthus's theory of population."[50] Darwin tilted Malthus's economic law concerning human populations and resources toward the world of nature and concluded that the same forces were at work in both realms. By this very slight maneuver, Darwin assured himself a place in history along with Plato, St. Thomas, and the other great natural philosophers.

First off, Darwin acknowledged the basic difference between artificial breeding and mating in nature. In artificial breeding, animals are selected in advance for mating based on the desire of the breeder to pass on certain desirable characteristics of the parents to the offspring. In the state of nature, no such planned parenthood exists. In the wild, mating is much more a matter of chance, circumstance, and opportunity. The result is random variations in the offspring, meaning the expression of traits that are totally devoid of any intention. Reproduction in nature, then, is like a giant lottery, meaning there is simply no intelligent goal in mind when two animals decide to commit themselves to each other for a brief, passing moment.

While Darwin was well aware that the process of mating in nature was random, he believed that the process of survival and reproduction was not. As to where chance leaves off and determinism sets in, Darwin argued that it's all a matter of the ability of the offspring to survive over his or her competition and produce more offspring, assuring that its particular traits will be passed on. To be more specific, while the appearance of new variations is randomly produced, those traits that prove in any way more "useful" in helping to assure the survival of one organism over another will likely be retained simply because the fittest organism will produce more offspring exhibiting those same traits. Silvan Schweber sums up Darwin's view of natural selection:

> Members of a species exhibit variations. . . . Some of these variations are heritable. Because in each generation more individuals are produced than can survive to reproduce, there is a struggle for existence. In this struggle . . . certain heritable traits will render an organism better adapted to its environment than other members of the species (not endowed with the trait or endowed with other traits). The fitter individuals—that is, the ones better adapted to their environment—will therefore leave more offspring (with similar traits).[51]

Stripped, then, of its pretensions, Darwin's case rested on the rather anthropocentric supposition that "the best-adapted biological forms were seen as surviving the struggle for life in the wild, in exactly the same way that the fittest individuals were thought to survive the rigours of industrialisation in *laissez-faire* Britain."[52]

Progress Through Struggle

Every era rests its fortunes on a few easily recitable aphorisms. For the Victorian gentleman it was the age of progress, and those in positions of power and influence never seemed to tire of hearing that magical expression evoked over and over again in their daily round. Certainly a civilization's cosmology ought to evoke a like-minded spirit, and Darwin's construction lived up to the expectations of the times. Darwin argued that "improvement" is the linchpin of the entire evolutionary process.

> When you contrast natural selection and "improvement," you seem always to overlook ... that every step in the natural selection of each species implies improvement in that species or relation to its condition of life. ... Improvement implies, I suppose, each form obtaining many parts or organs, all excellently adapted for their functions. As each species is improved, and as the number of forms will have increased, if we look to the whole course of time, the organic condition of life for other forms will become more complex, and there will be a necessity for other forms to become improved, or they will be exterminated: and I can see no limit to this process of improvement. . . .[53]

The idea of "no limit" to the process of improvement was in no way an invention of Darwin's. The prospect of unlimited improvement had been ensconced as the centerpiece of the Industrial Age from the time the French aristocrat Marquis de Condorcet had proclaimed:

> No bounds have been fixed to the improvement of the human faculties ... the perfectability of man is absolutely indefinite ... the progress of this perfectability, henceforth above the control of every power that would impede it, has no other limit than the duration of the globe upon which nature has placed us.[54]

Darwin looked into nature and saw the same forces at work there as Condorcet and others saw at work in Europe. In reassessing Darwinism one hundred years after his death, political commentator Tom Bethell suggests that "what Darwin really discovered was nothing more than the Victorian propensity to believe in progress."[55]

Progress to be sure. But at a price. There was no doubt in Darwin's mind that, in the very long run, natural selection confirmed the idea of steady improvement. But in the very short run, nature appeared to be a veritable jungle of hostility, danger, and violent activity. Darwin reasoned that any organism that was able to triumph over the forces threatening it from every side, successfully passing its inheritance on to its offspring, must certainly be said to contribute to the "improvement" of the species as a whole. Darwin made his feelings clear on this matter: "[There is] . . . one general law, leading to the advancement of all organic beings, namely, multiply, vary, let the strongest live and the weakest die."[56] It's not surprising that Darwin might reason this way, since this was precisely the view that the average Victorian gentleman held about himself in those middle decades of the nineteenth century.

It was in South America and the Galápagos Islands that Darwin formed his early impressions of a hostile nature at work. On his return home, fresh with the imagery of that scene in mind, Darwin could not help but make comparisons between what he had witnessed in the remote, desolate regions of South America and life as it existed in the streets of London. As Donald Worster suggests: "Undoubtedly his London life seemed to confirm what he had seen in South America and the Galápagos."[57]

The Galápagos were not very attractive places. Another visitor, Herman Melville, likened them to Hades. "In no world but a fallen one," he wrote, "could such lands exist."[58] Both Melville and Darwin were shocked at the fierceness of the environment. It was indeed foreboding. Danger lurked behind every shadow. "Darwin saw great masses of vultures and condors wheeling in the skies; and vampire bats, jaguars, and snakes now infested his mind."[59] It was a savage, primeval scene, menacing in every detail. Everywhere there was bloodletting, and the ferocious, unremittent battle for survival. The air was dank and foul and the thick stench of volcanic ash veiled the islands with a kind of ghoulish drape. Of the terrain, Darwin wrote that it was "a bro-

ken field of black basaltic lava, thrown into the most rugged
waves, and crossed by great fissures."[60] Darwin later reminisced
that the potholes, mounds, and ash of the Galápagos landscape
resembled "those parts of Staffordshire, where the great iron
foundries are most numerous."[61] The comparisons didn't stop
there. The images used by popular writers to describe Victorian
England bore strong resemblance to Darwin's description of life
on the Galápagos. England was rife with conflict. Everywhere
the naked eye fixed, one could see evidence of a furious struggle
to survive. Around every corner, there were eyes anxiously lying
in wait, ready to pounce, to trample, to snatch away and tear off
anything that was unable to resist their grasp. The new industrial landscape seemed grossly distorted. It was marked by jag-
ged mechanical protrusions whose steely mouths spewed forth a
film of gray residue that lingered over the cities, marking the
domain of the new industrial order.

The imagery of one island fused into the imagery of the other.
Everywhere Darwin gazed in the Galápagos, he saw England,
and everywhere his eyes took him in England, he saw the
Galápagos. Malthus's law of population brought the two experiences into a single vision. Natural selection provided the unifying focus. "The language of the *Origin*," says Barry Gale, "was
the language of the Victorian age . . . [with] its message of competition, its vision of conflict and struggle, its images of war and
destruction. . . ."[62]

Nature and the Bourgeoisie

Success in the struggle for survival was measured by the transmission of one's traits to one's offspring. For Darwin, inheritance
was central to natural evolution as assuredly as it was central to
the class whose values he so accurately reflected. Darwin created
an entire cosmology around inheritance, mediated by natural selection, and in its every detail, it reads like a running commen-
tary on the bourgeoisie's frame of mind in Victorian England.

To begin with, Darwin retained the idea of a hierarchy in nature; but unlike the ideas of earlier cosmologies, in his concept of nature creatures did not have their "status" bestowed on them from a higher authority. Instead, each organism in nature's hierarchy had to work hard to eke out its place in the natural scheme of things. Darwin's nature was the first meritocracy. It fit well with the temperament of a rising middle class that prided itself on being self-made. Darwin argued that only through the slow, tedious, trial-and-error process of constant struggle did organisms "earn" their place in nature's hierarchy. And unlike earlier cosmologies, in which everyone's place was fixed and unchangeable, Darwin's "evolutionary" schema made room for the continued biological advancement of one's heirs in nature's expanding hierarchy, thus providing still another engaging feature that was sure to find a supportive audience among the middle class.

The bourgeoisie were "collectors." More than any other group in history, they enjoyed nothing so much as acquisition. Even today, when one thinks of a bourgeois mentality, the notion of acquisition immediately leaps to mind. The idea of acquisition finds its likeness in Darwin's theory of inherited traits. Darwin's organisms passed on bodily acquisitions (their traits) to their offspring, thus assuring them a proper "inheritance" to give them a running start over their competition. Darwin's class passed on their material acquisitions to their offspring, assuring the same end.

Of course, even those considerations pale in contrast to the ultimate expectation Darwin fulfilled in his cosmology. As in every other concept of nature, Darwin offered up the prospect of everlasting life—no small feat, since his theory had removed God and heaven entirely from the picture. In the place of God, Darwin substituted natural selection; and in exchange for immortality in the next world, Darwin offered a modicum of immortality here in the earthly world. Darwin expressed quite a bit of concern over the question of immortality.

Believing as I do that man in the distant future will be a far more perfect creature than he now is, it is an intolerable thought that he and all other sentient beings are doomed to complete annihila-

tion after such long-continued slow progress. To those who fully
admit the immortality of the human soul, the destruction of our
world will not appear so dreadful.[63]

But Darwin was not one of those who believed in the immortality of the human soul. He was too much the scientist for that.
So he invented the next best thing—physical immortality of
sorts, through natural selection. Those able to triumph in the
struggle for survival were guaranteed a bit of immortality, as
their physical characteristics would continue to live on in their offspring, and since there is no end to the evolutionary process, they
could experience everlasting physical life vicariously through
their heirs and their heirs' heirs. As for those who were unsuccess-
ful in the battle to reproduce, they would be denied immortality.
Their physical death would mark the end of their line.

Nature's Economy

Over the past two decades, many historians of science have
taken another look at Darwin, his theory, and the personalities
and forces that influenced him. Writing in *The Journal of the History of Biology*, Silvan Schweber of Brandeis University sums up
the opinion of many of his colleagues when he notes that "Darwin's *Origin of Species* can be characterized as evolutionary
thought joining hands with British political economy and British philosophy of science"[64] in the mid-years of the nineteenth
century. Darwin borrowed heavily from the popular economic
thinking of the day. While, by Darwin's own admission,
Malthus's economic writings were a key influence in the development of his theory, Darwin was equally influenced by one of
the other great economic philosophers of the eighteenth century,
Adam Smith. An examination of Smith's and Darwin's writings
shows how deeply indebted the latter was to the thoughts Smith
penned in *The Wealth of Nations,* published in 1776.

It was Smith's theory of the division of labor that sparked
much of Darwin's thinking on evolutionary development and
the transmutation of species. According to Smith, "The greatest

improvement in the productive powers of labour, and the greater part of the skill, dexterity and judgement with which it is anywhere directed, or applied, seem to have been the effects of the division of labour."[65] In his famous pin factory example, Smith argued that whereas ten pin makers working on their own could produce a total of only two hundred pins, by dividing up the job into separate tasks, the same ten pin makers could produce 48,000 pins. Thus, through the proper exploitation of specialized tasks, it was possible to greatly expand production. Smith's argument was no mere idle speculation. It was, instead, just a rather erudite description of processes already at work in the economic life of the fledgling industrial power. The skilled craftsman responsible for the manufacturing of an entire product was being replaced by an industrial system in which each worker was in charge of a minute portion of the production process. At the Great Exhibition of 1851, Prince Albert praised "the great principle of the division of labour" as "the moving power of civilization."[66] This fundamental change in people's relationship to production helped to reshape humanity's own thoughts about the relationships and interworkings in nature itself. It was Darwin who was ultimately responsible for formalizing the analogy between economics and biology; and in so doing, says sociologist Robert Young, he gave "the mark of scientific respectability to the equation of the division of labor with the laws of life."

Darwin came to his theory by studying the works of biologist Henri Milne-Edwards, who, in turn, had based his observations of nature on the works of Adam Smith. In his *Élémens de Zoologie,* the first volume of which was published in 1834, Milne-Edwards set out to extend Smith's concept of the division of labor into the plant and animal world. Milne-Edwards first looked at the simplest of organisms, "whose faculties are most limited," and argued that "the interior of these organisms can be compared to a workshop where all the workers are employed in the execution of similar labors, and where consequently their number influence the quantity but not the nature of the products."[67] It is obvious, argued Milne-Edwards, that "as one rises in the series of beings, as one comes nearer to man, one sees organization becoming more complicated; the body of each animal becomes composed of parts which are more and more dissimilar to one another."[68] Milne-Edwards concludes that

> little-by-little the diverse functions localize themselves, and they all acquire instruments that are proper to themselves.[69] . . . [T]he greater the extent of the specialization of functions and the division of labor, the more the number of different parts and the complication of the machine has to increase . . . the number of dissimilar parts that make up an organism and the magnitude of the differences that these parts present are the indices of the degree to which the division of labor has been carried out.[70]

There was no doubt in Milne-Edwards's mind that "the principle which seems to have guided nature in the perfectibility of beings is, as one sees, precisely one of those which have had the greatest influence on the progress of human industr[ial] technology: *the division of labor.*"[71]

Darwin contends that he gained his insight into the divergence of species and the division of labor in nature from reading Milne-Edwards. Silvan Schweber argues that Darwin might indeed be fudging, since the works of Adam Smith and the other economists of the day were all very familiar to him as well. Schweber contends that attributing his insight to Milne-Edwards allowed Darwin "to ascribe the principles of the physiological division of labor to an eminent zoologist and philosopher of biology, rather than to the political economists."[72] Schweber says that the obfuscation was deliberate and intended to temper any charges of rank anthropocentrism by his critics, not to mention the possible charge of acting in a quite unscientific manner.

Darwin's intentions, however, matter less than the analogy itself. As Schweber points out, Darwin acknowledges that his "explanation of the divergence of characters was essentially equivalent to the concept of the 'physiological division of labor' that Milne-Edwards had promulgated,"[73] and the latter, in turn, "credited his formulation of the concept of the division of physiological labor to the writings of political economists,"[74] most notably Adam Smith, the grand old father of the idea.

The idea of a division of labor provided a solution to a question that had long plagued Darwin: exactly why there was a "tendency in organic beings descended from the same stock to diverge in character as they become modified."[75] The answer, according to Darwin, was to be found in the idea that "the same spot will support more life if occupied by very diverse forms."[76] Darwin argued that in nature, as in Adam Smith's pin factory and the English economy,

> the greatest number of organic beings (or more strictly the greatest amount of life) can be supported on any area, by the greatest amount of their diversification. . . .[77] For in any country, a far greater number of individuals descended from the same parents can be supported, when greatly modified in many different ways, in habits, constitution and structure, so as to fill as many places, as possible, in the polity of nature, than when not at all or only slightly modified.[78]

As Young observes, by finding in nature the same kind of division of labor at work as that found in the English factory system, Darwin provided "a scientific guarantee of the rightness of the property and work relations of industrial society."[79] It's no wonder Darwin was so enthusiastically embraced by the entrepreneurs of the period. This was a time of great labor unrest. There was talk of syndicalism in the streets. New workingmen's organizations were pressing for labor reforms, including better working conditions, better wages, and a shorter work week. The tension between the capitalist factory owners and the working class was ever-present, threatening to ignite into open confrontation at the slightest provocation. The bourgeoisie was in need of a "proper" justification for the new factory system with its dehumanizing process of division of labor. By claiming that a similar process was at work in nature, Darwin provided an ideal rationale for those capitalists hell-bent on holding the line against any fundamental challenge to the economic hierarchy they managed and profited from.

Darwin's concept of divergence also provided an ideal defense for English imperialism during the heyday of its colonial expansion. By the time Darwin had published *The Origin of Species,* the

English could rightfully claim that "the sun never sets on the British Empire." Britain was at its peak of influence. This tiny island off the European coast had successfully extended its tentacles to the far recesses of the globe. British bureaucrats, laborers, soldiers, and entrepreneurs were busily engaged in other people's lives, lands, and businesses, converting much of the world to the British frame of mind.

While some justification for colonial policies might be found in the proposition that the English were bringing civilization and religion to the savage backwaters of history, it was not enough to gloss over the rather obvious signs of avarice and greed that colored Britain's imperialist ambitions. The British bourgeoisie was in need of a far more impressive argument, one that was airtight, incontrovertible, beyond question. They found the argument they were looking for in Darwin's theory of divergence. For his part, Darwin found the analogy he was looking for from British economic practices and particularly the practice of colonization. So, once again, the economic forces of the period provided the metaphor; and Darwin, in turn, provided the rationalization by projecting that metaphor directly into nature. It should be emphasized, once again, that the entire process unfolded rather unconsciously.

Darwin argued that "all organic beings are striving to seize on each place in the economy of nature."[80] The ensuing struggle creates a kind of "chronic labor surplus."[81] That is, there are always likely to be too many organisms competing for too few niches in nature. Because most economic roles are already taken, there is a constant and savage contest among nature's creatures to either hold on to the economic niche they have or to cast someone else out of theirs. Darwin's explanation of the surplus labor problem in nature would no doubt have elicited more than a smirk of recognition on the faces of most English workers of the period. The 1830s and 1840s—the decades Darwin was fashioning his theory—were marked by chronic unemployment. The bitter struggle for a limited number of jobs resulted in widespread hostility and confrontation as Englishmen increasingly turned against their neighbors.

Darwin goes on to say that there are only two ways to promote an organism's survival: Either compete for the existing ecological niches or find new ones that have yet to be filled. Again, using the idea of divergence, Darwin argued that occasionally a new organism will exhibit new traits sufficiently different from its peers to allow it to fill a previously unoccupied niche in nature. Divergence, then, was nature's way of laying "the potential to devise new ways of living."[82] Divergent traits provided "an innate resourcefulness that could make use of resources whose very existence had not been suspected."[83] Migration into new niches lessens the competition for existing slots and at the same time opens up entirely new areas for exploitation. For example, "the evolution of a new kind of grass may create a series of new niches for animals still undeveloped, and they in turn may someday serve as prey for new species of predators."[84] As Donald Worster points out in *Nature's Economy,* diversity for Darwin "was nature's way of getting round the fiercely competitive struggle for limited resources."[85] For the millions of Englishmen forced to leave the British Isles in the nineteenth century to look for new economic opportunities in alien lands, Darwin's notion of divergence made more than a little sense. In the colonies, there were, as yet, untapped opportunities, economic niches ready to be filled and exploited. In contrast to the niggardly supply of economic possibilities that presented themselves on the depressed English homefront in the 1830s and 1840s, divergence seemed a welcome reprieve.

Darwin found in nature ample biological justification for British colonization. In fact, he was anything but subtle on this score. He wrote that "when two races of men meet, they act precisely like two species of animals—they fight, eat each other, bring diseases to each other, etc; but then comes the more deadly struggle, namely which have the best fitted organisation, or instinct (i.e., intellect in man) to gain the day."[86] On numerous occasions Darwin made known his belief that natural selection was proving so effective that it would be only a matter of time before "endless numbers of the lower races will have been eliminated by the higher civilized races throughout the world."[87] At a time when Britain was extending her influence into the remote regions of the globe, colonizing new lands and people, it was most reassuring to know that wherever the Union Jack was raised, natural selection was being allowed to flourish.

It was not surprising, given Darwin's views on the matter of race superiority and natural selection, that he would enthusiastically endorse the ideas of his cousin Francis Galton, who proposed the science of eugenics, by which he meant the systematic breeding of people "to further the ends of evolution more rapidly and with less distress than if events were left to their own course."[88] Darwin thought that his cousin's proposal was a "grand" idea; and his only reservation, says Gertrude Himmelfarb, was the concern that "few people could be counted on to cooperate intelligently."[89] In his book *The Descent of Man*, Darwin is even more adamant when he considers the contrast between artificial breeding and man's own haphazard way of going about mating.

> The weak members of civilised societies propagate their kind. No one who has attended to the breeding of domestic animals will doubt that this must be highly injurious to the race of man. It is surprising how soon a want of care, or care wrongly directed, leads to the degeneration of a domestic race; but excepting in the case of man himself, hardly any one is so ignorant as to allow his worst animals to breed.[90]

Hand in hand with the idea of progress, material acquisition, inheritance, division of labor, and colonization was the idea of individualism. The bourgeois era distinguished itself in philosophy and outlook by an obsessive concern with the rights, needs, aspirations, and drives of the individual. While in previous epochs life was viewed more in terms of the collective eyes of the family, tribe, clan, or community, the advent of the Industrial Age brought with it a marked shift in focus to the individual. Just as Newton had imagined a material world made up solely of autonomous pieces of matter interacting against one another, the economic life of industrial England increasingly came to reflect a similar organizing style. The political and economic dogma of the day extolled the virtues of individualism, and everywhere one turned there was great support for the notion of unfettering the individual from the chains that bound him to a larger whole, so that he might be free to pursue his own interests and inclinations.

The idea of the individual as the primary unit of economic activity was first given concrete expression in the works of Adam Smith. His account of the relationship of the individual to the economic system was duplicated almost in its entirety in Darwin's *Origin of Species*. Smith starts with the assumption that the individual, being the primary economic agent, should be given free rein to maximize his own material interests. As to the argument that such unrestrained activity might, on occasion, prove harmful to the larger community, Smith vigorously disagrees. Economic historian John McCulloch captures Smith's philosophy when he asserts that it is only ". . . by the spontaneous and unconstrained . . . efforts of individuals to improve their condition . . . and by them only, that nations become rich and powerful."[91] Smith acknowledges that such behavior is supremely self-serving and self-centered, but contends that the very act of individual selfishness benefits the general well-being of others. In words that have since been canonized by the architects and prime movers of industrial capitalism, McCulloch eloquently summarized Smith's philosophy by observing that

> every individual is constantly exerting himself to find out the most advantageous methods of employing his capital and labour. It is true that it is his own advantage, and not that of the society, which he has in view; but a society being nothing more than a collection of individuals, it is plain that each, in steadily pursuing his own aggrandisement, is following that precise line of conduct which is most for the public advantage.[92]

But what is to prevent the entire experience from descending into a Hobbesian nightmare where each is pitted against all and where the pursuit of individual interest ends up in perpetual warfare and ultimately chaos? According to Smith, the competition between individuals is mediated by "an invisible hand," a law of nature that regulates supply and demand in the economic marketplace with complete reliability. The invisible hand works very simply. When demand for a commodity or service goes up, the suppliers raise their price accordingly to take advantage of the situation. When the price becomes too high, demand will slacken off or move to some other commodity or service, forcing the sellers to lower the price to the point where demand is rekindled, and so on. As long as no encumbrances are put in the way of the free exercise of individual interests on the part of the buyers and sellers, the invisible hand will ensure that a proper balance will always be maintained between production and consumption. This balancing mechanism, then, works like a well-calibrated pendulum, swinging back and forth, constantly adjusting population demands with resource limitations. Smith was ready to admit that the invisible hand could not guarantee against individual failures, nor should it. In fact, that was the price paid for a system in which someone's loss was always someone else's gain. In the end, society as a whole would be the winner, as the success of the few would advance the economic fortunes of the rest of the community.

Darwin's ideas about how nature operated were virtually identical to the economic formulation of an "invisible hand." To begin with, Darwin's theory of biological evolution centered on the fortunes of the individual organism. He agreed with both Smith and Malthus that in nature as in society each individual is absorbed with maximizing its own self-interest and surviving in the struggle with others over limited resources: "Each individual of each species holds its place by its own struggle and capacity of acquiring nourishment."[93] The problem for Darwin was to try to understand how such individual activity meshed into a working whole. The key was to be found in the parallel relationship between Smith's invisible hand, Malthus's law of population and resources, and his own theory of natural selection. Darwin reasoned that just as an external law is constantly at work in the economic sphere, regulating and balancing supply and demand, a similar law—natural selection—must be constantly at work in nature, forever regulating and balancing the supply of resources against the demand for those resources.

Likewise, whereas each individual organism is interested only in its own survival, its triumph can't help but advance the common good, since its traits live on in its offspring, thus assuring a never-ending process of gradual improvement in the biological characteristics of the species as a whole. While nature produces many failures, they are the price paid for the evolutionary advancement of life itself. Peter J. Bowles captures the close correlation between the invisible hand and Darwin's theory of evolution. He observes that "both the balance of nature and the *laissez-faire* view of competition were based on the belief that nature and society are fundamentally harmonious systems in which apparent conflict serves for the benefit of all."[94]

Of particular interest is the novel way Darwin's theory accommodated the age-old schism between humanity's desire for self-containment on the one hand and its sense of mutual obligation and relationship on the other. Darwin's nature was like an extended family, in which all species in the hierarchy were biologically related. This familial relationship imposed an overwhelming obligation in that humanity had to acknowledge that it "owed" its entire biological makeup to the sacrifices made in the evolutionary struggle from the tiniest life form onward. If *Homo sapiens* was merely the sum total of all the adaptive characteristics passed on from generation to generation over eons of time, it was also at the apex of the evolutionary hierarchy; and its superior status evoked a sense of resentment in being related to and obligated to inferior biological specimens. Humanity seeks to disassociate itself from its dependence on other living things, and so it is caught in the age-old trap of seeking self-containment while acknowledging its inexorable relationship to all other forms of life.

Darwin's theory offered a resolution to humanity's perennial crisis of guilt. By proposing that each organism's drive for self-containment actually benefited the species as well as nature as a whole, Darwin found a convenient formula for expiating the accumulating guilt of an age when self-interest and personal aggrandizement ruled supreme. There could be nothing wrong in the desire for total self-sufficiency, even if it meant the ruthless exploitation of one's fellow human beings and other creatures, since the common good was always advanced in the process. If an organism benefited at the expense of others during its lifetime, at least it gave back a gift to nature in the form of its superior traits, which were passed on to its offspring, assuring the continued evolutionary amelioration of all of nature. Thus, while Darwin was adamant in his defense of the individual as the most important single element of the system, he nonetheless created a theory which expressed, perhaps better than any other in history, the close biological relationship that exists between all living things.

Darwin's theory was a welcome tonic for the bourgeoisie. It reinforced the idea of an invisible hand guiding the marketplace and seemed to confirm the positions taken by the gentry and middle class on the key social issues of the day. Of course this shouldn't be very surprising since Darwin's own ideas were heavily "influenced" by the political attitudes of the same class which later embraced his theory so enthusiastically. In fact, at the very time that Darwin was piecing together his theory, a great debate was being waged in British politics over social reforms, particularly the issue of relief to the poor. Nowhere were the political biases of his class more vociferously expressed than in the great reform debate of the 1830s, and Darwin was, no doubt, understandably influenced by the opinion of his peers. Most of the controversy centered on the passage of the Poor Law Act. This piece of legislation was the subject of endless discussion in the drawing rooms and private clubs of the bourgeois class. More than any other domestic issue, it symbolized the political orientation of the emerging middle class. It is instructive to note that the act was passed just four years before Darwin articulated his theory of natural selection, and the attitude expressed in it was faithfully reproduced in his own piece of legislation.

At issue was society's responsibility to the poor. It was common practice for the local parishes to supplement low wages with public allotment, the amount determined by the size of the family. As Smith's invisible hand and Malthus's law of population gained an increasing number of followers among the gentry and middle class, they began to argue that the parish laws were violating the laws of nature by rewarding the indolent and least fit, thus allowing the poor to reproduce their kind, to the detriment of both nature and the social order. Armed with a new code of political ethics steeped in so-called natural law, the bourgeois class argued that virtually all forms of public charity should be withdrawn (except for the aged or infirm) and that the poor should be put to work. That's exactly what happened. Under the new legislation, workhouses were set up, and all able-bodied souls, including entire families, were sent there to labor under the most appalling conditions for what amounted to slave wages.

The new attitude toward the poor was a combination of punishment and benign neglect. The gentry and the middle class believed that if the laws of nature were left to work without interference, the struggle for survival would act as the supreme arbiter, weeding out and eliminating the least fit in society and ensuring a healthy environment where only the fit would flourish. Herbert Spencer voiced the feelings of the bourgeois class when he argued:

> The well-being of existing humanity, and the unfolding of it into this ultimate perfection, are both secured by that same beneficent, though severe discipline, to which the animate creation at large is subject; a discipline which is pitiless in the working out of good; a felicity-pursuing law which never swerves for the avoidance of partial and temporary suffering. The poverty of the incapable, the distresses that come upon the imprudent, the starvation of the idle, and those shoulderings aside of the weak by the strong, which leave so many "in shallows and in miseries," are the decrees of a large far-seeing benevolence.[95]

Over and over again, Darwin heard such arguments expressed by his contemporaries. Not surprisingly, they found their way into his conception of how things work in nature. Subsequently, his own theory made its way back to those same drawing rooms and private clubs where it, in turn, was used to reconfirm the very arguments that had inspired his own thoughts on the question of the proper workings of nature.

Though it was never his intention, Darwin's theory became the chief weapon in the political arsenal of the new bourgeois

class. Its central tenets reinforced the idea that "all attempts to
reform social processes were efforts to remedy the irremediable,
that they interfered with the wisdom of nature, that they could
lead only to degeneration."[96] In place of legislative reform, the
Social Darwinists argued for the slow trial-and-error process of
gradual change regulated by natural selection and the invisible
hand. In short, evolution, not revolution. As historian Richard
Hofstadter observes:

> The idea of development over aeons brought new force to another
> familiar idea in conservative political theory, the conception that
> all sound development must be slow and unhurried. Society could
> be envisaged as an organism . . . which could change only at the
> glacial pace at which new species are produced in nature.[97]

With nature squarely on its side, the new bourgeois class was
equipped with the biological rationale it needed to cement its
own political position. At the same time, Darwin's theory of
evolution provided it with a sophisticated defense against the
working class and poor, who were demanding radical social reform and a measure of relief from the dire economic straits they
found themselves in.

Nature as Machine

Darwin's description of the evolution of species bears a remarkable resemblance to the workings of the industrial production
process in which machines were assembled from their individual
parts. While it would be grossly unfair to suggest that Darwin
knowingly borrowed the concept of industrial assembly, his theory of biological evolution did indeed reflect a similar method of
production in nature. Each new species was seen as an assem-
blage of individual parts organized into new combinations and
arrangements and with additional improvements designed to
increase both their complexity and their efficiency.

It's difficult to fault Darwin for relying so heavily on machine imagery. There was simply no way to escape the overwhelming presence of the machine in English life at the time. Here was this marvelous new technology that was reshaping the world. "Naturally" everyone was anxious to extend its application to every facet of life. One could hardly expect biologists to remain aloof from the excitement of the day.

Well before Darwin began to borrow machine imagery, it was already well established as the central metaphor in physics. Descartes, Newton, and the other early architects of modern science had already come to the conclusion that the universe itself resembled a well-organized machine that ran effortlessly and perpetually by way of precise mathematical laws. The whole universal machine was composed of "elementary building blocks" assembled in such a way as to ensure the proper functioning of the entire mechanism. In this new scheme of things God was transformed from a craftsman deity to a clockmaker. He was congratulated on being the supreme mathematician who had engineered the whole plan and set it in motion. Eventually he was retired from the scene altogether, as succeeding generations became more and more intoxicated with the power this newfound paradigm provided them with.

The mechanical world view dealt exclusively with physical phenomena that could be mathematically measured. By separating and then eliminating all the qualities of life from the quantities of which they were a part, the architects of the machine world view were left with a cold, inert universe made up entirely of dead matter. This mechanical conception of physical existence laid the groundwork for a thorough desacralization of all forms of life during the ensuing Industrial Age. As historian Neal Gillespie of Georgia State University points out, "Positive science, conceived as a system of empirically verifiable facts and processes and the theories linking them, required the radical desacralization of nature."[98] So convinced were the scientists of the truth of their mechanical world view that they readily agreed with René Descartes's claim that living creatures were merely soulless automata, no more than simple machines totally unable to experience either pain or pleasure. This attitude led to bar-

baric experiments on animals in the scientific laboratories in an effort to better understand what made these machines tick. Jean de La Fontaine's account of such experiments demonstrates the extent to which the new mechanical cosmology had succeeded in desacralizing the last vestiges of aliveness in nature.

> They administered beatings to dogs with perfect indifference and made fun of those who pitied the creatures as if they had felt pain. They said that the animals were clocks; that the cries they emitted when struck, were only the noise of a little spring which had been touched, but that the whole body was without feeling. They nailed poor animals up on boards by their four paws to vivisect them and see the circulation of the blood which was a great subject of conversation.[99]

Darwin liked to think of himself as the Newton of biology, and for good reason. Taking up where these earlier scientists had left off, Darwin successfully transformed the idea of a mechanical universe into a mechanical theory of the origin and development of species. Nature was viewed as a workshop well stocked with a variety of detachable, replaceable parts that were constantly rearranging themselves into more and more complex and efficient living machines. Before the age of the machine, living creatures were viewed as "wholes." This idea fit well with an artisan mode of production in which the craftsman "molded" his creation from a primordial substance. In the craft form of production each mold exists as a separate unit. Products are made whole, and each is rigidly constrained by the mold it is stamped from. Thus, the long-standing view of nature as a collection of fixed creations, each one molded exactly as the original without any variation whatsoever, was quite congenial with the craft form of production. This traditional view of nature was overthrown and replaced with a radically new conception, compatible with the new form of industrial production. Darwin came to view living things as the sum total of lifeless, inanimate parts "assembled" together in various functional combinations. Darwin admitted that it was no longer possible for him to even imagine that living creatures were created whole and in their entirety. With the idea of step-by-step machine assembly so firmly implanted in the British mind, Darwin concluded:

> Almost every part of every organic being is so beautifully related to its complex conditions of life that it seems as improbable that any part should have been suddenly produced perfect, as that a complex machine should have been invented by man in a perfect state.[100]

Darwin turned living organisms into machinelike objects. In desacralizing nature so thoroughly, Darwin and his contemporaries severed whatever slim strands of animism still remained in people's cosmological visions. Donald Worster captures the significance of the accomplishment.

> By reducing plants and animals to insensate matter, mere conglomerates of atomic particles devoid of internal purpose or intelligence, the naturalist was removing the remaining barriers to unrestrained economic exploitation.[101]

In defense of Darwin, it can be said that his thinking was in no way sinister. It was not malice that led him to deaden nature so convincingly. If anything, he was undoubtedly unconscious of how his mind was being influenced by the environment it was set down in. In this regard, Darwin was not alone. As John Michell, the author of *The View over Atlantis,* so accurately points out, in every age

> people justify their activities . . . by making a cosmology that reflects their own particular ideas and obsessions. . . . The mechanical universe idea was temporarily convenient to an age which was concerned with the development of mechanical inventions.[102]

Evolutionary Ethics

Like other cosmologists through history, Darwin dealt with the age-old question of morality in such a manner that would ensure that good coincided with the prevailing mode of organizing nature and evil with activity that undermined the objectives of the society.

In his own secular way Darwin saw nature as fallen. It was a place full of evil intention. Darwin was aghast at the suffering inflicted in nature. "There seems to me too much misery in the world."[103] The idea of a merciless and vengeful nature greatly troubled this mild-mannered English gentleman. According to Himmelfarb, Darwin reconciled his conflicting sentiments by seeing "the cruelty of nature as part of a larger beneficent design."[104] His theory of natural selection served much like the scales of justice, meting out punishments and rewards with greater impartiality than one might expect of Solomon himself.

> It may be said that natural selection is daily and hourly scrutinizing, throughout the world, every variation, even the slightest; rejecting that which is bad, preserving and adding up all that is good; silently and insensibly working, whenever and wherever opportunity offers, at the improvement of each organic being.[105]

Preserving the good and rejecting the bad: precisely what every code of ethics demands of its adherents. As to the fateful question of what is good, Darwin answers with all the combined authority of the utilitarian ethos of his day. Good is what is useful, or to put it in more of a biological frame, good is what is adaptive. Whatever traits best advance one's prospect for survival must be good. In contrast, to cite one of Darwin's more brash contemporaries, Herbert Spencer, "The root of all evil is the 'non-adaptation of constitution to conditions.' "[106] Whether one uses "useful" or "adaptive," the intent is the same. Good behavior, that is, useful or adaptive behavior, is that behavior which in some way enhances the ability of the organism to maximize its self-sufficiency, to minimize its dependency.

When Darwin talks of evolution as a process of weeding out imperfections and of perpetual reorganization to higher and higher levels of perfectibility, he is unconsciously talking about the desire to reach an ultimate state of total self-containment. Of course a species that reached such an exalted state would bear a striking resemblance to what previous cosmologists thought God to be. God has always been associated with the idea of total self-containment. To be God is to be without need, to be totally self-sufficient and invulnerable.

Before the Industrial Age people aspired to be more like God, and since God has always stood for self-containment, it only made sense that people would associate good behavior or godlike behavior with self-sufficient behavior.

While "self-sufficiency" and "being good" have always been intertwined, they have been open to widely different interpretations through history. For the Greeks, being good and being self-sufficient meant being materially secure. For the early Christians, self-sufficiency and good behavior meant the opposite. To be selfless, without earthly wants or worldly needs, was for the Christian regarded as the ultimate expression of self-containment and good behavior. Regardless of interpretation, people have consistently sought to "perfect" themselves, to live a life that reflected, as best it could, God's own goodness. But since God didn't figure in Darwin's scheme, he needed a substitute. Without God to aspire to, what, indeed, motivated the climb up the ladder of evolution?

As industrialism gradually began to take hold, succeeding generations increasingly turned their faces from the pearly gates to the factory gates, where a substitute for God was found waiting eagerly to enlist their unswerving faith and devotion. This new secular god was called efficiency, and it served exactly the same function for the people of science and industry as earlier divinities had for their own times.

The idea of efficiency motivated Darwin's theory of evolution just as forcefully as it motivated industrial life in nineteenth-century England. Silvan Schweber points out that "Darwin characterizes the largest favorable areas in which the most dominant animals evolve as the 'most efficient workshop.' "[107] It's not surprising that Darwin should think this way. After all, England had managed to transform itself into the single-minded pursuit of increased efficiency. It was the new god and was treated with all the respect that an earlier Englishman might have heaped on the Savior himself.

Indeed, efficiency embodies many of the same attributes that one might ascribe to God. Efficiency is the attempt to maximize returns by minimizing the expenditure of energy. Perfect efficiency would amount to having everything at one's disposal that could possibly be produced without having to exert any energy whatsoever. In previous epochs, humanity cast God in that role. In lieu of God, Darwin conjured up a natural world where each succeeding species was more efficient than its predecessor; that is, better able to maximize the return on its investment of time and energy. Thus evolution was always advancing toward the perfectly efficient organism, meaning the perfectly self-contained organism, meaning an organism invulnerable to all outside influences, meaning an organism remarkably similar in constitution to God.

Perfect efficiency bore a striking resemblance to God in still another way. Efficiency is an attempt to minimize physical decay. Each time we expend energy, part of us dies. Therefore, the less energy one gives up and the more energy one gets in return, the better able one is to preserve oneself against the ravages of time. If an organism never had to expend any energy at all and only took energy in, it could live forever. Greater efficiency, then, means cheating time and death. It is humanity's way of overcoming the limitations of time and space, of becoming immortal, like God himself, who, after all, is the most efficient of all, since he expends no energy, yet everything does his bidding.

Efficiency is humanity's way of striving for self-sufficiency. The less one has to give up in order to get something back in return, the less dependent one is. As in past cosmologies, where the good is a measure of self-sufficiency, modern men and women have come to believe that the more efficient they become, the better they and the world are as a consequence. Increased efficiency is the highest value of industrial society. In the age of the machine, being good means being efficient. Donald Worster sums up the mood of the age and the impact it had on the new cosmological formulations of Darwin and his colleagues:

While scientists busied themselves in collecting and classifying the facts of nature ... they also managed to create an ecological model that accurately mirrored the popular bourgeois mood. Its fundamental assumption was that the "economy" of nature is designed by Providence to maximize production and efficiency.[108]

Politics and Darwinism

Darwin's theory of evolution followed the steam engine across the globe. Everywhere industrialism took hold, evolution dogma soon became a permanent fixture. It attracted widespread support across geographic boundaries because it seemed to explain the nature of things in terms that were easily recognizable. Bewildered by the changes sweeping over their lives, people were anxious for some kind of grand explanation that could put everything into focus. Darwin's theory met the test, precisely because it was able to find in nature the same forces at work as people were experiencing in their day-to-day lives in the factories and towns that pioneered the Industrial Age.

Darwin's theory was not a cosmology just for capitalist England, even though it was nursed on British soil. It was, instead, a cosmology suited to explain and justify the Industrial Age itself; and because of its broader appeal, it was taken up and embraced by most of the major ideological schools of the period. As Geoffrey West notes:

> Darwinism has been seized upon by all parties as a strong bulwark in defence of their contradictory preconceptions. On the one hand Nietzsche, on the other Marx, and between them most shades of Aristocracy, Democracy, Individualism, Socialism, Capitalism, Militarism, Materialism and even Religion.[109]

Marx and Engels were quick to seize on Darwin's theory for their own ideological ends. Marx said that Darwin's book serves as "a basis in natural science for the class struggle in history."[110] In fact, he was so excited about the implications of evolutionary theory for socialism that he asked Darwin's permission to dedicate *Das Kapital* to him, but the English scientist kindly rejected

the offer for fear that his association with the treatise might undermine the credibility of his own work among his peers. Marx
valued two things above all else in Darwin's work. First, the theory of evolution succeeded in delivering "the death-blow . . . for
the first time to teleology in the natural sciences."[111] According
to Himmelfarb, Marx recognized that, besides removing God effectively from the picture, Darwin was challenging the "abstract
materialism of most natural sciences"[112] by introducing the historical process into science. Marx and Engels fervently believed
that Darwin's own theory of the origin and development of species paralleled their own thoughts on the origin and development of culture. At Marx's funeral Engels proclaimed: "Just as
Darwin discovered the law of evolution in organic nature, so
Marx discovered the law of evolution in human history."[113]

In her book *Darwin and the Darwinian Revolution,* Gertrude
Himmelfarb points out that Darwin, Marx, and Engels were all
convinced that the universe runs by "fixed laws" and that "God
was as powerless as individual men to interfere with the internal,
self-adjusting dialectic of change and development."[114] To be
sure, there were also parts of Darwin's theory that these socialists
took exception to. For example, while Darwin thought that the
struggle for existence was a "permanent condition," Marx and
Engels were of the belief that it characterized evolution only up
through the bourgeois era and would fade away with the dawn
of the classless society.

In a letter of 1875, Engels wrote a friend in what—given the
source—some might consider an ironic vein. It is clear that he
perceived that what Darwin was doing was projecting the industrial life of Britain onto nature and then using nature as a justification for British industrial life. His insight bears repeating.

> The whole Darwinist teaching of the struggle for existence is sim-
> ply a transference from society to nature of Hobbes' doctrine of
> *bellum omnium contra omnes* and of the bourgeois economic doctrine
> of competition, together with Malthus' theory of population.
> When the conjurer's trick has been performed . . . the same
> theories are transferred back again from organic nature to history

and it is now claimed that their validity as eternal laws of human society has been proved.[115]

One would expect such an insight to result in the complete and utter denunciation of Darwin's theory as a bourgeois trick, but it didn't. The reason it didn't is that neither Engels nor Marx nor any of the other architects of the Industrial Age was anxious to throw out a theory that, despite its particulars, provided the best cosmological defense of industrialism available. So Engels overlooked some of the specific details as well as the inspiration behind the theory and, like his contemporaries, cheerfully embraced Darwin's explanation of nature as the unimpeachable truth of the matter.

The great world historian Oswald Spengler placed the theory of evolution in its most succinct context with the simple observation that Darwin's entire thesis amounted to little more than "the application of economics to biology."[116] So it did, but if the application to nature was motivated by economics, the transference of nature's laws back to society was motivated by politics. With the publication of the *Origin,* the bourgeois class could rationalize its economic behavior by appealing to the universal laws of nature as its ultimate authority. It was possible, indeed acceptable, to justify the brutal exploitation of the working poor and imperialist adventures abroad all in the name of faithfully obeying the "laws of nature." Gertrude Himmelfarb surveys the political repercussions of what has come to be known as Social Darwinism and concludes that it has served as the undisputed centerpiece for the politics of the Industrial Age. As a political instrument, Darwinism has exalted

> competition, power and violence over convention, ethics, and religion. Thus it has become a portmanteau of nationalism, imperialism, militarism, and dictatorship, of the cults of the hero, the superman, and the master race.[117]

Up to now, scholars have been more than willing to admit that Darwin's theory of evolution has been exploited over and over again for political purposes. What they have been less willing to see is that the theory itself is little more than a projection onto nature of people's own economic activity with the intent of finding a political justification in nature for the way we treat each other and our environment. In other words, equally important to how Darwin's theory was used was how it was conceived. Darwin's theory reflected the historical moment in its conception as well as in its application. What Darwin discovered was not so much nature as it is but the workings of nineteenth-century industrialism, which he then unconsciously projected onto nature. In doing so, he provided a biological legitimization for the politics of his class during the early stages of the bourgeois era.

Darwin's theory succeeded precisely because it fit so well the temper of the times. People readily identified with Darwin's view of how nature was organized because its basic tenets conformed so nicely with the way society was organizing itself and its own immediate environment. Darwin constructed a theory of nature that, in its every particular, reinforced the operating assumptions of the industrial order. In so doing, he provided something much more valuable than a mere theory of nature. Darwin gave industrial man and woman the assurance they needed to prevail against any nagging doubts they might have regarding the correctness of their behavior. His theory confirmed what they so anxiously wanted to believe: that the way they were organizing their existence was indeed "harmonious" with the natural order of things.

Darwin's cosmology sanctioned an entire age of history. Convinced that their own behavior was in consort with the workings of nature, industrial man and woman were armed with the ultimate justification they needed to continue their relentless exploitation of the environment and their fellow human beings without ever having to stop for even a moment to reflect on the consequences of their actions.

PART FOUR

THE DARWINIAN SUNSET

The Passing of a Paradigm

Intellectuals love to disagree. It comes with the job. Put a representative sample of academics together in a large conference room and chances are assured that they will be ready and eager to take exception to almost any proposition put forth for discussion. Scholars are trained to qualify, quibble, banter, and berate at the drop of a hat. A good scholar prides himself on the ability to maintain a modicum of healthy skepticism, regardless of the issue at hand.

In the intellectual world it is considered bad form to hold any proposition beyond the pale of rigorous criticism and debate. No matter of concern to the human mind is considered resolved and therefore closed to further question. Well, almost. The fact is, there is one proposition that has been largely spared from the constant carping that surrounds most intellectual questions. It seems that Darwin's theory of evolution has been granted a reprieve of sorts. For decades it has enjoyed a rather privileged position within the academic community. Many would be quick to defend its special status. After all, were it not for the theory of evolution, we would be without a context for discussing all the other important issues concerning life. The way we look at the world, the way we frame the important questions, the approaches we choose to address the issues of the day, all need a frame of reference beyond reproach. The theory of evolution is the unimpeachable starting point. It is our world view, and without it everything else is unintelligible. The great twentieth-century scientist-philosopher Sir Julian Huxley understood exactly how important evolutionary theory is to explaining everything else when he argued:

> It is essential for evolution to become the central core of any educational system, because it is evolution, in the broad sense, that links inorganic nature with life, and the stars with the earth, and matter with mind, and animals with man. Human history is a continuation of biological evolution in a different form.[1]

Huxley's hopes have been answered. Evolutionary theory has been enshrined as the centerpiece of our educational system, and elaborate walls have been erected around it to protect it from unnecessary abuse. Great care is taken to ensure that it is not damaged, for even the smallest rupture could seriously call into question the entire intellectual foundation of the modern world view. If a survey were taken, no doubt virtually all our academic leaders would find themselves in agreement with biologist Julian Huxley, who proclaimed, with more than a bit of bravado, that "Darwin's theory . . . is no longer a theory but a fact. No serious scientist would deny the fact that evolution has occurred just as he would not deny that the earth goes around the sun."[2]

In previous times, to question the prevailing orthodoxy was to invite charges of heresy. Now that science has eclipsed theology as the dominant paradigm, we prefer to replace theologically inspired insults with more clinically acceptable terms. Today, says biologist Garrett Hardin, if a member of the academic establishment dares to question Darwin's theory of evolution, he "inevitably attracts the speculative psychiatric eye to himself."[3] In other words, his sanity is immediately questioned.

One person undaunted by the prospect of such scrutiny is Dr.

Colin Patterson. He is a senior paleontologist at the British Natural History Museum, in London. Dr. Patterson is the author of the book *Evolution* and is recognized as the world's leading paleoichthyologist. On November 5, 1981, Dr. Patterson delivered a speech before a group of experts on evolutionary theory at the American Museum of Natural History. Dr. Patterson dared to suggest to his colleagues that the scientific theory that he and they had devoted a lifetime to was mere speculation, without any significant evidence to back it up. Here's how Dr. Patterson explained his change of mind concerning the theory of evolution:

> Last year I had a sudden realization. For over twenty years I had thought I was working on evolution in some way. One morning I woke up and something had happened in the night; and it struck me that I had been working on this stuff for twenty years and there was not one thing I knew about it. That's quite a shock, to learn that one can be so misled so long. . . . So for the last few weeks I've tried putting a simple question to various people and groups of people. . . . Can you tell me anything you know about evolution, any one thing, any one thing that is true? . . . All I got . . . was silence. . . .
>
> The absence of answers seems to suggest that . . . evolution does not convey any knowledge, or, if so, I haven't yet heard it. . . . I think many people in this room would acknowledge that during the last few years, if you had thought about it at all, you have experienced a shift from evolution as knowledge to evolution as faith. I know that it's true of me and I think it is true of a good many of you here. . . . Evolution not only conveys no knowledge but seems somehow to convey antiknowledge.[4]

Imagine that! A scientist dispensing with the cosmology of the age with such utter abandon. Surely the man is mad. Is it really possible to believe that we are all wrong—that we have been living under a grand illusion no more real than Alice's Wonderland? Psychiatrist Karl Stern of the University of Montreal says it is quite possible indeed. As to the question of sanity vs. insanity, Stern asks us all to detach ourselves from our preconceived biases and consider the merits of the Darwinian argument. The theory, says Stern, goes something like this:

> At a certain moment of time, the temperature of the Earth was such that it became most favourable for the aggregation of carbon atoms and oxygen with the nitrogen-hydrogen combination, and that from random occurrences of large clusters molecules occurred which were most favourably structured for the coming about of life, and from that point it went on through vast stretches of time, until through processes of natural selection a being finally occurred which is capable of choosing love over hate and justice over injustice, of writing poetry like that of Dante, composing music like that of Mozart, and making drawings like those of Leonardo.[5]

Stern's opinion of the evolutionary theory is not likely to win many friends within the scientific community. Speaking strictly from the point of view of a psychiatrist, he argues:

> Such a view of cosmogenesis is crazy. And I do not at all mean crazy in the sense of slangy invective but rather in the technical meaning of psychotic. Indeed such a view has much in common with certain aspects of schizophrenic thinking.[6]

Stern and Patterson are not alone. While biology teachers continue to teach the most up-to-date textbook version of Darwin's theory of evolution to the children of the 1980s, some of the high priests of biology have all but abandoned their own sacred texts. Although unwilling to claim that evolution per se is a crazy idea, many of them are more than prepared to commit Darwin's version of it to the historical archives. Remarkably little has been written in the popular press about this rebellion in the making. The *coup d'état* has unfolded rather quietly within the semi-sequestered domain of official academic conferences and scholarly journals. The first inkling that things were not well with Darwinism came, interestingly enough, during the centennial celebration of Darwin's theory held at the University of

Chicago in 1959. One of the speakers, paleontologist Everett Claire Olson of the University of California, let it be known that

> there exists, as well, a generally silent group of students engaged in biological pursuits who tend to disagree with much of the current thought, but say and write little because they are not particularly interested, do not see that controversy over evolution is of any particular importance, or are so strongly in disagreement that it seems futile to undertake the monumental task of controverting the immense body of information and theory that exists in the formulation of modern thinking.[7]

As to how many had actually deserted ranks, Olson contended that it is "difficult to judge the size and composition of this silent segment, but there is no doubt that the numbers are not inconsiderable."[8]

Since then, the silence has been breached. One by one, the dissenters have surfaced. Their voices, once a faint murmur, have swelled into a chorus of discontent.

Right now an intense struggle is going on within the profession, pitting the dyed-in-the-wool Darwinists against a new generation of theoreticians who are anxiously casting around for a more satisfactory explanation of the origin and development of species. The battle recently extended directly into London's Natural History Museum, long considered a bulwark of Darwinian thinking. At issue was a pamphlet published by the museum which qualified Darwinism by saying, "If the theory of Evolution is true."[9] "If" indeed! Much of the scientific community was aghast. To even suggest such a possibility—and coming from the British Natural History Museum—was enough to steam the bifocals of many a don at Cambridge, Oxford, Sussex, and other esteemed institutions throughout the kingdom. An editorial appearing in *Nature*, the unofficial voice of the scientific establishment, rebuked museum officials in no uncertain terms. Noting that "most scientists would rather lose their right hand than begin a sentence with 'if the theory of Evolution is true,'" the editorial asked rhetorically, "What purpose except confusion can be served by these weasel words?"[10]

Other establishment bastions have been caught up in the debate. For example, many years ago, G. A. Kerkut, professor of Physiology and Biochemistry at the University of Southampton, England, published a book critical of Darwin's theory entitled *Implications of Evolution.* Kerkut concluded: "The attempt to explain all living forms in terms of an evolution *from a unique source,* though a brave and valid attempt, is one that is premature and not satisfactorily supported by present-day evidence."[11] An unusually candid review of the book appearing in the *American Scientist,* the official publication of the prestigious Sigma Xi scientific fraternity, acknowledged what many had long suspected but were afraid to entertain, especially in print. Speaking to the book as well as to Darwin's theory, the review stated:

> This is a book with a disturbing message; it points to some unseemly cracks in the foundations. One is disturbed because what is said gives us the uneasy feeling that we knew it for a long time deep down but were never willing to admit this even to ourselves. . . . The particular truth is simply that we have no reliable evidence as to the evolutionary sequence . . . one can find qualified, professional arguments for any group being the descendant of almost any other. . . . We have all been telling our students for years not to accept any statement on its face value but to examine the evidence, and, therefore, it is rather a shock to discover we have failed to follow our own sound advice.[12]

More recently, another book critical of Darwin's theory was published in France, by Dr. Pierre P. Grassé. The book, *Evolution of Living Organisms,* greatly intensified the debate at hand because of the eminence of the source. Dr. Grassé is one of the world's greatest living biologists. In his review of the book, Theodosius Dobzhansky, a member of the old guard and a staunch defender of Darwinist theory, had to admit that Grassé's observations were impossible to ignore simply because of his vast research experience. Grassé is the editor of the twenty-eight volumes of *Traité de Zoologie* and ex-president of the French Academy of Sciences. According to Dobzhansky, "His knowledge of the living world is encyclopedic."[13] After decades of careful scholarship, Grassé concluded simply:

Their success among certain biologists, philosophers, and sociologists notwithstanding, the explanatory doctrines of biological evolution do not stand up to an objective, in-depth criticism. They prove to be either in conflict with reality or else incapable of solving the major problems involved.[14]

Much to the consternation of his colleagues, Grassé ends up by issuing the single most devastating indictment that can ever be leveled against a field that professes to be scientific.

Through use and abuse of hidden postulates, of bold, often ill-founded extrapolations, a pseudoscience has been created. It is taking root in the very heart of biology and is leading astray many biochemists and biologists, who sincerely believe that the accuracy of fundamental concepts has been demonstrated, which is not the case.[15]

The charge that evolution theory is a "pseudoscience" is being heard with increasing frequency. In the introduction to a 1971 edition of Darwin's *Origin of Species* British zoologist Leonard Matthews expressed the concern of many of his colleagues when he said: "The fact of evolution is the backbone of biology, and biology is thus in the peculiar position of being a science founded on an unproved theory—is it then a science or faith?"[16]

To qualify as science, Darwin's theory should be provable by means of the scientific method. In other words, its hypothesis should be capable of being tested experimentally. The scientific method involves a three-step process of verification. First, data or phenomena are observed. Second, a hypothesis is formulated, based on the observations. The hypothesis allows the scientists to make certain predictions regarding the data. Third, experiments are set up to test the hypothesis and to determine whether the predictions are valid. If the outcome of the experiments validates the prediction, the hypothesis is considered verified. At the heart of the scientific method is experimental repeatability or reproducibility. According to biologist George Gaylord Simpson, one of the leading Darwinists of the twentieth century,

The important distinction between science and those other systematizations (i.e., the arts, philosophy and theology) is that science is self-testing and self-correcting. The testing and correcting are done by means of observations that can be repeated with essentially the same results.[17]

By the very standards scientists choose to limit what may properly fall under the purview of their discipline, the theory of evolution does not qualify. Even Theodosius Dobzhansky admits that theories of evolution evade the scientific process. Dobzhansky laments the fact that "evolutionary happenings are unique, unrepeatable, and irreversible. It is as impossible to turn a vertebrate into a fish as it is to effect the reverse transformation."[18] Dobzhansky is chagrined and reluctantly acknowledges that "the applicability of the experimental method to the study of such unique historical processes is severely restricted before all else by the time intervals involved, which far exceed the lifetime of any human experimenter."[19]

Embarrassing, to say the least. Here is a body of thought, incapable of being scientifically tested, claiming to be scientific in nature. It can't be observed, it can't be reproduced, it can't be tested, and yet its proponents demand that it be regarded as the supreme, unimpeachable truth regarding the origin and development of life! One might think that any self-regarding scientist would want to know where the proof is. The great Russian biochemist Aleksandr Oparin says that if it's proof we're after, we're never going to find it: "Proof in the sense in which one thinks of it in chemistry and physics is not attainable in the problem of primordial biogenesis."[20]

If we can't prove evolution by the scientific method, at least we can't disprove it either. Of course, that's true, but the same thing might be said of every other speculation that is not subject to the rigors of the scientific method. In order to fall within the scientific domain, a theory must be falsifiable. In other words, it must be capable of being subjected to experiment to prove whether it is true or false. For example, Newtonian physics is falsifiable. Experiments can be conducted to prove whether or not Newton's laws are true. However, evolution theory, like the belief in God, is not falsifiable. There is no way to either prove or disprove its claim scientifically. Even Darwin understood as much. In a letter written in 1863, he admitted:

> When we descend to details, we can prove that no species has changed (i.e. we cannot prove that a single species has changed); nor can we prove that the supposed changes are beneficial, which is the groundwork of the theory.[21]

If not based on scientific observation, then evolution must be a matter of personal faith. About the best that can be said about the theory is that it represents a belief that many people share about how life developed, a belief that can be neither proved nor disproved. Of course, everyone is entitled to his own beliefs, speculations, and personal convictions, but evolution proponents profess that their theory represents much more than a simple article of faith. It is pure truth, they contend, even though unprovable, and in their zeal they are unwilling to brook any opposition to its central tenets. Writing in the introduction to still another publication of Darwin's *Origin,* entomologist W. R. Thompson reproached the "defenders of the faith" for their unscientific conduct.

> This situation, where men rally to the defense of a doctrine they are unable to define scientifically, much less demonstrate with scientific rigor, attempting to maintain its credit with the public by the suppression of criticism and the elimination of difficulties, is abnormal and undesirable in science.[22]

The ferocity with which many evolutionists set forth their views and the high level of intolerance they demonstrate toward alternative frames of reference should be cause enough for pause, if not alarm. In their behavior, one senses a familiar pattern, one that has been with us since the first time man began to formulate a cosmology. The evolutionist today is every bit the "true believer." Baptized in the theory of natural selection, he is prepared to spread the good news and bring his fellow human beings to accept Darwin's teachings. Edwin G. Conklin, late professor of Biology at Princeton University, recognized the per-

vasive sense of religiosity that permeated the thinking of his colleagues when he remarked:

> The concept of organic evolution is very highly prized by biolo-
> gists, for many of whom it is an object of genuinely religious de-
> votion, because they regard it as a supreme integrative
> principle.[23]

Conklin goes on to say that this religious fervor "is probably the
reason why severe methodological criticism employed in other
departments of biology has not yet been brought to bear on evolutionary speculation."[24]

Long able to fend off criticism from outside the profession, now
the traditionalists find themselves confronted, for the first time,
with serious opposition among their own kind. The field of biology is undergoing a reformation as potentially profound as the
one that shook the Christian world four centuries ago. Like the
early Protestants, a new generation of "protesters" is emerging
within the scientific community. These reformers believe in evolution but not in the explanation of it handed down by the high
priests of the faith. Remaining faithful to the idea of evolution,
they are determined to save the doctrine by radically changing
the liturgy.

In 1980, molecular biologists, embryologists, ecologists, and
paleontologists from around the world gathered at Chicago's
Field Museum to discuss the theory of evolution. The conference
quickly turned into a battleground, pitting the prevailing orthodoxy against the reformers. By the end of the sessions it became
obvious to many of the delegates and observers attending that a
historic shift in thinking had taken place regarding the theory of
evolution. Had Darwin been there, it is likely that he would not
have been able to muster up more than a flagging handful of
votes for his particular version of the origin and development of
species.

The debate centered on what is known as "the modern synthesis," the updated version of Darwinist theory that has dominated

the field for more than forty years. The modern synthesis brought Darwin together with the developments in genetics in the early part of the twentieth century. Remember, Darwin had not developed an adequate theory concerning heredity. That had to await Mendel's discovery in 1865 regarding the transmission of heredity information to the offspring. New discoveries in the field of genetics in the 1920s and 1930s finally led to what is known as the "synthetic theory" (or the neo-Darwinian synthesis). The theory was advanced in a series of books written in the 1930s and 1940s by Theodosius Dobzhansky, Julian Huxley, Ledyard Stebbins, Gaylord Simpson, Ernst Mayr, and others. According to Mayr, "The new 'synthetic theory' of evolution amplified Darwin's theory."[25] Mayr asserts that, like Darwin's original theory, the new synthesis is

> characterized by the complete rejection of the inheritance of acquired characters, an emphasis on the gradualness of evolution, the realization that evolutionary phenomena are population phenomena and a reaffirmation of the overwhelming importance of natural selection.[26]

While the neo-Darwinian synthesis retained the essential Darwin, it added genes to the equation, thus providing a biological explanation for the actual transfer of hereditary information from parent to offspring. Mayr explains natural selection dressed up in genes as a two-step process.

> The first step is the production (through recombination, mutation, and chance events) of genetic variability. . . . Most of the variation produced . . . is random in that it is not caused by, and is unrelated to, the current needs of the organism or the nature of its environment. The second step of natural selection, selection itself, is an extrinsic ordering principle. In a population of thousands or millions of unique individuals some will have sets of genes that are better suited to the currently prevailing assortment of ecological pressures. Such individuals will have a statistically greater probability of surviving and of leaving survivors than other members of the population. It is this second step in natural selection that determines evolutionary direction, increasing the frequency

of genes and constellations of genes that are adaptive at a given time and place, increasing fitness, promoting specialization and giving rise to adaptive radiation and to what may be loosely described as evolutionary progress.[27]

The new synthesis resurrected Darwin, making him contemporary with the advances in genetics during this century. Sir Julian Huxley fondly referred to the miracle of transformation as "this reborn Darwinism, this mutated phoenix risen from the ashes of the pyre."[28] This giant bird hovered over the field of biology for four long decades before being finally shot down at the Field Museum Conference.

The main question on everyone's mind at the forum was whether or not genetic variations within a population (microevolution) are also responsible for producing new species (macroevolution). In other words, was the neo-Darwinian synthesis correct in its assertion that, over time, all of the many small genetic mutations expressed by individuals within a species have a cumulative effect (as a result of natural selection) eventually resulting in the metamorphosis or transformation into an entirely new, novel species? According to *Science* magazine, "The answer can be given as a clear, NO."[29] Stephen Jay Gould, a Harvard paleontologist, and one of the leaders in the reformation movement, summed up the new consensus:

> I have been reluctant to admit it . . . but if Mayr's characterization of the synthetic theory is accurate, then that theory, as a general proposition, is effectively dead, despite its persistence as textbook orthodoxy.[30]

The Fossil Record

The neo-Darwinian synthesis has been given the death sentence; and, ironically, the evidence used to convict it was the very evidence first bandied about by its supporters in defense of the theory. It is the fossil record that now figures as a major argument in the case against Darwinism. The rocks tell their own story and, in almost every detail, what they have to say contradicts the story put forth by the Darwinists and neo-Darwinists regarding the origin and development of life on this planet.

The only evidence that exists concerning the past history of life on earth is found in fossils embedded in rock formations. Without the fossils, it would be virtually impossible to make even the most halfhearted conjecture as to how life might have unfolded. Grassé reminds his fellow scientists that

> the process of evolution is revealed only through fossil forms. A knowledge of paleontology is, therefore, a prerequisite; only paleontology can provide . . . the evidence of evolution and reveal its course or mechanisms.[31]

Every schoolchild is introduced to Darwin's theory of evolution by way of the fossils. Page after page of textbook photos showing actual fossils and colorful re-creations of fossils is presented as concrete proof that we do indeed have a past, and a rather long one at that. The evidence is lined up in a neat, orderly progression, showing the unmistakable step-by-step, gradual evolution of life from the tiniest living creature to man himself. Everything is so tidy. Nothing appears out of place, and one is struck with how well everything seems to fit.

For generations of youngsters initiated into the passages of life by way of the fossils, it is difficult to even dare to suggest that the entire exercise is little more than an imaginary trip through time. Yet that happens to be the case. Evolution, as depicted in the textbooks, is simply how some people "imagine" that life might have unfolded. If it is hard to accept that such is the case, it is probably because we have all been so thoroughly indoctrinated in this particular story of how life unfolded that we now have an investment of sorts to protect. After all, we gave our time over to studying all the pictures and charts; we memorized the categories and classifications; we spent long hours preparing for examinations on the subject. So our immediate inclination is to dismiss the possibility of error as pure balderdash.

To ensure that future generations of youngsters are spared our humiliation, Dr. Grassé suggests that an epigraph be attached "to every book on evolution,"[32] to read as follows:

> From the almost total absence of fossil evidence relative to the origin of the phyla, it follows that any explanation of the mechanism in the creative evolution of the fundamental structural plans is heavily burdened with hypotheses. ... The lack of direct evidence leads to the formulation of pure conjectures as to the genesis of the phyla; we do not even have a basis to determine the extent to which these opinions are correct.[33]

The fact is, the fossil record was never considered very good evidence for Darwin's theory. Even Darwin recognized this sorry state of affairs, and he said as much; but chances are that not one person in a million has ever heard his reservations on the matter. After contrasting his own theory, which postulated that each species slowly evolves through a host of "intermediary forms" into novel new species, with the evidence contained in the fossil record, Darwin asked rhetorically, Where are the "intermediate links"? The answer was nowhere to be found in the geologic record. As Darwin candidly admits: "Geology assuredly does not reveal any such finely graduated organic change; and this, perhaps, is the most obvious and gravest objection which can be urged against any theory [of evolution]."[34]

Darwin's chapter on the fossil record was entitled "On the Imperfections of the Geologic Record." His choice of words proved to be prophetic. By every reasonable standard, the geologic records should have been teeming with intermediate varieties and transitional forms between species. Yet Darwin was unable to find any convincing examples. Rather than admit defeat, he argued that all the evidence was not yet in ... that not enough fossils had been looked at to make a definitive judgment. Darwin hoped that over time, intermediate forms would be discovered, thus validating his hypothesis. He was not to be so honored. In the one-hundred-plus years since Darwin asked for more time, millions of fossils have been collected and analyzed. The results have been less than complimentary to Darwin. According to David Raup, curator of the Field Museum in Chicago, where examples of 20 percent of all known fossil species are kept, the evidence does not in any way support Darwin's contention of a gradual, step-by-step evolution with countless intermediate forms linking one species to another.

> Most people assume that fossils provide a very important part of the general argument made in favor of Darwinian interpretations of the history of life. . . . Well, we are now about 120 years after Darwin, and knowledge of the fossil record has been greatly expanded. . . . Ironically, we have *even fewer examples* of evolutionary transition than we had in Darwin's time.[35]

Frankly, Dr. Raup is being unnecessarily kind. What the "record" shows is nearly a century of fudging and finagling by scientists attempting to force various fossil morsels and fragments to conform with Darwin's notions, all to no avail. Today the millions of fossils stand as very visible, ever-present reminders of the paltriness of the arguments and the overall shabbiness of the theory that marches under the banner of evolution.

A journey through the geologic record should convince even the skeptic of the utter nonsense of gradual evolutionary transformation. The first sign that fact and theory are at odds is found in the oldest rock formations where fossils are discovered.

Fossils first show up in the geologic record in highly complex forms. Mollusks, jellyfish, sponges, crustaceans, and the rest of the invertebrates are all found together in what paleontologists conveniently label as the Cambrian Period. Here is where the mystery begins. It is assumed by experts in the field that it would have taken at least a billion years for life to have evolved to even this level of complexity. That being so, large numbers of their fossil ancestors should be lodged in the pre-Cambrian rock formations. The problem is, not so much as a single "multicellular" ancestral fossil has ever been found in the earlier rocks. Paleobotanist Daniel Axelrod of the University of California states the problem clearly and unequivocally:

> One of the major unsolved problems of geology and evolution is the occurrence of diversified, multicellular marine invertebrates

in Lower Cambrian rocks on all the continents and their absence
in rocks of greater age. . . . However, when we turn to examine the
Pre-Cambrian rocks for the forerunners of these Early Cambrian
fossils, they are nowhere to be found.[36]

From an evolutionary point of view, this makes absolutely no
sense. The pre-Cambrian fossil record should be brimming over
with all the ancestors of the multicellular invertebrates that are
peppered through the Cambrian rocks. Remember, the theory of
evolution is based on the notion that life developed from nonlife
and proceeded through gradual mutations, from the very simplest forms of life to the more complex. Why, then, does the geologic record show only very advanced, complex life forms?
Where are the earlier stages? Gaylord Simpson, among others,
confesses that the absence of intermediate forms of life that
might bridge the gap between simple microorganisms and the
complex invertebrates of the Cambrian Era is the "major mystery of the history of life."[37]

Actually, the discrepancy between theory and fact only begins
with the Cambrian period. It turns out that our paleontologists
have been unable to find a single convincing intermediate or
transitional form linking the various life forms in the fossil
record. For example, the experts estimate the time period between the Cambrian Era and the so-called Ordovician Era,
when the first animal-like fossils show up in the rocks, to be
nearly 100 million years. According to the British scientist F. D.
Ommaney, an internationally known expert on fish life: "How
this earliest chordate stock evolved, what stages of development
it went through to eventually give rise to truly fish-like creatures,
we do not know."[38] The reason they don't know is that they have
been unable to find any intermediate forms of life during that
100-million-year period to link the fishlike creatures with their
alleged ancestors of the Cambrian Era.

Then there's the question of the fish. Once again, their origin
and ancestry are shrouded in mystery. Paleontologist Alfred
Romer of Harvard's Museum of Comparative Zoology observes:

In sediments of late Silurian and early Devonian age, numerous
fishlike vertebrates of varied types are present, and it is obvious

that a long evolutionary history had taken place before that time. But of that history we are mainly ignorant.[39]

Why the fishes' evolutionary history should be "obvious" when all the evidence, by Romer's own account, precludes just such a history is testimony to the hold evolutionary doctrine enjoys over scientists whose own research contradicts the theory. Perhaps more to the point on the matter is Errol White, formerly the Keeper of the Department of Paleontology of the British Natural History Museum. In his presidential speech to the prestigious Linnaean Society of London, White, who is an authority on lungfish, informed his colleagues, "Whatever ideas authorities may have on the subject, the lungfishes, like every other major group of fishes that I know, have their origins firmly based in nothing."[40]

And so it goes, with the paleontologists convinced that simpler species evolve into more complex species and then, at each level of the fossil record, candidly acknowledging that there is virtually no evidence whatsoever of the actual "evolution" taking place.

For example, consider the transformation from fish to amphibians. The structural differences between the two are so great that it would have taken millions of years of gradual evolutionary changes, in which time countless intermediate forms would have had to emerge, in order to link fish with amphibians. Yet, as biochemist Duane Gish points out, the links are nowhere to be found. The record shows that between the fin of the crossopterygian and the foot of the amphibian *Ichthyostega* is an anatomical gap so large that it begs the question once again: Where are the millions of intermediate forms that would be required to exist in order for the former to evolve into the latter?

There is a basic difference in anatomy between all fishes and all amphibians not bridged by transitional forms.

In all fishes, living or fossil, the pelvic bones are small and loosely embedded in muscle. There is no connection between the pelvic bones and the vertebral column. None is needed. The pelvic bones do not and could not support the weight of the body. . . . In tetrapod amphibians, living or fossil, on the other hand, the pelvic bones are very large and firmly attached to the vertebral column. This is the type of anatomy an animal must have to walk. It is the type of anatomy found in all living or fossil tetrapod amphibians but which is absent in all living or fossil fishes. There are no transitional forms.[41]

Nowhere are the shortcomings of evolution theory more pronounced than in the case of birds. One of the world's experts on the subject, William Elgin Swinton, is forced to admit that once again "there is no fossil evidence of the stages through which the remarkable change from reptile to bird was achieved."[42] Of course, there was a time when paleontologists thought that they had located at least one potential candidate that might, with the appropriate stretch of the imagination, qualify as an intermediate form. It was called *Archaeopteryx* and was much heralded. Even though, in every important respect, it was definitely a bird—complete with wings, feathers, and the ability to fly—it enjoyed the unfortunate distinction of being endowed with certain rather unappealing reptilian-like features, including teeth, vertebrae along the tail, and tiny clawlike appendages running along the edge of the wings. While not very pretty to behold, it was a sight for sore eyes among the paleontologists, who were quick to embrace it as their one real shred of evidence linking reptiles to birds.

It should be pointed out that the reptilian-like features found in *Archaeopteryx* were more cosmetic than structural, and that since its discovery other still-living birds have been found that exhibit clawlike appendages, thus casting doubt over the inflated importance attached to this creature. But these other considerations are no longer really significant, because the case of *Archaeopteryx* was finally put to rest in 1977, when *Science News* reported that a bird fossil had been found in rocks from the same geological period, demonstrating that the so-called missing link lived and flew side by side with other birds, thus precluding the possibility of its being an ancient ancestor.[43] In fact, *Archaeopteryx* was just another bird, not a very handsome representative, to be sure, but still functionally very much a bird.

While *Archaeopteryx* is still likely to grace most biology textbooks with its toothy smile long after it has been dismissed by many paleontologists, there is still another creature that has traditionally been afforded an even more exalted status. Almost every introductory textbook contains the famous picture of the evolution of the horse. There it is: first the tiny *Eohippus* prancing through the glades, then getting larger, more sure-footed, and faster, and finally looking like the thoroughbred of today. On a recent television show over PBS entitled "Did Darwin Get It Wrong?" Darwin scholar Norman Macbeth finally exposed the great horse caper that had gone undetected for nearly eighty years. According to Macbeth:

> About 1905 an exhibit was set up [in the American Museum of Natural History] showing all these horses. . . . They were arranged in order of size. Everybody interpreted them as a genealogical series. But they are not a genealogical series; there is no descent among them. They were found at different times, in different places and they're merely arranged according to size.
>
> But it's impossible to get them out of the textbooks. . . . As a matter of fact, many of the biologists themselves forget what they are. I had a radio debate with a paleontologist some years ago and when I said there were no phylogenies, he told me I should go out to the Museum and look at the series of horses. I said, "But, Professor, they are not a family tree; they are just a collection of sizes." He said, "I forgot that."[44]

When it comes to the horse, says famed biologist Richard Goldschmidt, "the decisive steps are abrupt without transition."[45]

We could go on and on, but the point has already been made by the experts in the field. The world's leading paleontologists have been forced to admit that the fossil record simply does not bear out the Darwinian contention that there has been a gradual evolution over eons of time from the simple forms of life to the complex. David B. Kitts, professor of Geology at the University of Oklahoma, sums up the evidence against the theory when he observes: "Evolution requires intermediate forms between species and paleontology does not provide them."[46]

Stephen Jay Gould of Harvard and Niles Eldredge of the American Museum of Natural History put it even more bluntly: "Phyletic gradualism [gradual evolution] . . . was never seen in the rocks."[47] The two distinguished scientists go on to dismiss Darwin's notion of gradual evolutionary change as an expression of "the cultural and political biases of 19th century liberalism."[48] They fail to mention what cultural biases their own ideas are steeped in, but that's getting slightly ahead of the story.

Breeding and Species Stasis

Darwin based his theory of evolution on extrapolations from animal and plant breeding. As he observed the breeding techniques up close, he was impressed with the great number of varieties that could be obtained. Every species exhibits a great deal of variation. No one doubts for a moment that variation is a very real and ever-present reality, whether exhibited in controlled breeding experiments or in nature. There are countless varieties of apples and oranges. The question that has dogged Darwin, however, from the time he first advanced his thesis, is whether all the many variations within a species have a long-term cumulative effect, eventually resulting in the transformation to an entirely new and separate species. In other words, can one argue from apples to oranges? Darwin believed one could, even though, up to that time, no breeder had ever successfully performed such a feat. The breeders, for their part, did not share Darwin's optimism. Experience taught them that there were limits to what one could expect from breeding, and those limits were morphological. The point being, if you breed a dog over generations, you can come up with big and little dogs, fat and thin dogs, dogs with short tails and long tails, with curly hair or straight hair, but you will always get another dog and not a cat. Darwin recognized the problem but dismissed the hard evidence with his usual explanation that not enough time had elapsed for the many small variations to add up to major transformations.

Now, a hundred years later, the scientific community has become a bit more impatient with Darwin's plea for additional time. The developments in breeding techniques over the past half century have done little to confirm Darwin's prognosis and much to undermine the thinly veiled speculation Darwin was able to conjure up in support of his contention. As was the case with the fossil record, it appears that population breeding, Darwin's chief piece of evidence in support of his theory, turns out, in hindsight, to be the best piece of evidence against his claims. Loren Eiseley, one of the world's best-known anthropologists, captures the irony of the situation.

> It would appear that careful domestic breeding, whatever it may do to improve the quality of race horses or cabbages, is not actually in itself the road to the endless biological deviation which is evolution. There is great irony in this situation, for more than almost any other single factor, domestic breeding has been used as an argument for the reality of evolution.[49]

At the heart of the neo-Darwinian synthesis remains the idea that mutations within a species gradually accumulate over time, resulting in a metamorphosis to a brand-new species. This extrapolation from microchanges to macrochanges is what evolution is supposed to be all about. Yet the facts just don't support the theory. While breeders are apt to agree that a great deal of change can be brought about "within a species" by crossbreeding and selection, they would concur with University of Florida zoologist Edward S. Deevy's wry observation: "Wheat is still wheat, and not, for instance, grapefruit; and we can no more grow wings on pigs than hens can make cylindrical eggs."[50]

This rather obvious fact is understood by every professional breeder. Luther Burbank, perhaps the most famous breeder of the twentieth century, says that in all his years of practical experience, he came to realize that, in regard to breeding, there is an incessant law always at work that constrains and limits the extent of variations possible within a species.

I know from my experience that I can develop a plum half an inch
long or one 2½ inches long with every possible length in between,
but I am willing to admit that it is hopeless to try to get a plum
the size of a small pea, or one as big as a grapefruit. . . . In short,
there are limits to the development possible, and these limits follow a law. . . . [It is the law] of the reversion to the Average. . . .
Experiments carried on extensively have given us scientific proof
of what we had already guessed by observation; namely, that
plants and animals all tend to revert, in successive generations, toward a given mean or average. . . . In short, there is undoubtedly
a pull toward the mean which keeps all living things within some
more or less fixed limitations.[51]

What the scientists have learned about breeding goes a long
way toward solving a nagging problem that has plagued the theory of evolution from the beginning. That is, if evolution is constantly at work in nature, why is it that the geologic record
shows all the many species of plants and animals maintaining
themselves without change for millions of years? The geologic
record does not show evolutionary change, but the opposite.
Plants and animal species show up suddenly in the rocks and
then remain structurally the same for millions of years, until
they become extinct. Of course, over the countless generations of
a species' existence, mutations are occurring continually; but
there is not so much as a shred of evidence to suggest that such
mutations alter the species itself in any fundamental way.

Dr. Grassé poses the question this way: "How does the Darwinian mutational interpretation of evolution account for the
fact that the species that have been the most stable—some of
them for the last hundreds of millions of years—have mutated as
much as the others do?"[52] Grassé concludes: "Once one has noticed microvariations (on the one hand) and specific stability (on
the other), it seems very difficult to conclude that the former
(microvariation) comes into play in the evolutionary process."[53]
Grassé says that the evidence forces us "to deny any evolutionary
value whatever to the mutations we observe in the existing fauna
and flora."[54]

If anything, mutations appear to serve a stabilizing function, not a transforming one. Mutations guarantee that enough diversity or variety will be maintained within a population to ensure the species continued existence. Studies over the years have shown that when a species is bred to eliminate genetic diversity, the resulting homogeneous strains lack the variability necessary to promote their survival. In this respect, the evidence in breeding experiments leads to a conclusion that is exactly opposite to the one Darwin came up with. Darwin looked at artificial breeding and concluded that it was creating a more "fit" animal or plant, one better able to survive. Unfortunately, he confused more lucrative with more fit. While breeding techniques produce chickens with more eggs and cows with more milk and sheep with more wool and corn with bigger ears, they do so at the expense of the population's ability to survive. By selecting for certain traits and weeding out all the other characteristics deemed unuseful in an economic sense, the breeders produce a strain that is weakened and less resistant to harmful changes in the environment. According to Douglas Scott Falconer, formerly chairman of the Department of Genetics at the University of Edinburgh:

> The improvements that have been made by selection in these [domesticated breeds] have clearly been accompanied by a reduction of fitness for life under natural conditions, and only the fact that domesticated animals and plants do not live under natural conditions has allowed these improvements to be made.[55]

Artificial breeding, then, makes a population more profitable but less fit.

According to Grassé, mutations are "merely hereditary fluctuations around a medium position; a swing to the right, a swing to the left, but no final evolutionary effect . . . they modify what pre-exists."[56] Whereas Darwin thought that variations led to new species, the evidence proves the contrary: namely, that variation improves the ability of the species to maintain itself "against" radical change.

If there were any lingering doubt as to the incredible tenacity of species, experiments with the little fruit fly have done much to convince many respectable scientists that stasis, not transformation, is the general rule of nature. The fruit fly has long been the favorite object of mutation experiments because of its fast gestation period (twelve days). X rays have been used to increase the mutation rate in the fruit fly by 15,000 percent. All in all, scientists have been able to "catalyze the fruit fly evolutionary process such that what has been seen to occur in *Drosophila* (fruit fly) is the equivalent of many millions of years of normal mutations and evolution."[57] Even with this tremendous speedup of mutations, scientists have never been able to come up with anything other than another fruit fly. More important, what all these experiments demonstrate is that the fruit fly can vary within certain upper and lower limits but will never go beyond them. For example, Ernst Mayr reported on two experiments performed on the fruit fly back in 1948. In the first experiment, the fly was selected for a decrease in bristles and, in the second experiment, for an increase in bristles. Starting with a parent stock averaging 36 bristles, it was possible after thirty generations to lower the average to 25 bristles, "but then the line became sterile and died out."[58] In the second experiment, the average number of bristles was increased from 36 to 56; then sterility set in. Mayr concluded with the following observation:

> Obviously any drastic improvement under selection must seriously deplete the store of genetic variability. . . . The most frequent correlated response of one-sided selection is a drop in general fitness. This plagues virtually every breeding experiment.[59]

Bacteria offer still another compelling example of the stasis of species. Bacteria are the most prolific form of life. They account for 75 percent of all life forms, have allegedly been around for nearly 3 billion years, and if allowed to go unchecked for only thirty-six hours they would reproduce in numbers that would cover the entire planet "to a thickness of over a foot."[60] Bacteria produce more mutants than any other form of life; yet they have never reproduced anything other than another strain of bacteria.

Darwin observed variation within a species, then immediately jumped to the conclusion that it was responsible for the transformation of one species into another. The fossil record and modern breeding techniques both argue convincingly that Darwin and his twentieth-century apologists have been in error Both the fossils and experiments in breeding overwhelmingly attest to the fact that variations within a species promote stasis, not transformation; yet the extrapolation continues, despite all the concrete evidence to the contrary.

For example, take the case of the famous peppered moth, *Biston betularia*. Whenever someone questions evolution theory, proponents are quick to cite the miracle of the moth as ironclad proof of evolution at work, when in truth it proves the contrary. Back in 1924, zoologist H. B. D. Kettlewell of Oxford University demonstrated that in the industrial regions of England, darkened by pollution and soot, darker-colored moths proliferated over lighter-colored moths because they were able to go undetected by their natural enemies and survived longer to reproduce offspring. For over fifty years biologists have been parading the peppered moth past the reviewing stand at official academic conferences, and its stately presence has fluttered through the pages of many an introductory textbook on biology, always with the pronouncement that at last evolution has been proved. Needless to say, the biologists are a bit confused on the matter. As historian Gertrude Himmelfarb has pointed out, "No one questions the operation of natural selection on this level,"[61] but "by what right are we to extrapolate the pattern by which colour or other such superficial characters are governed, to the origin of species, let alone of orders, classes, phyla of living organisms."[62] All the peppered moth example proves is that variety in a species—in this case black-and-white moths—ensures that when environmental conditions radically change, the species itself may be able to adapt with sufficient success to maintain its own perpetuation as a species. Far from being an example of evolution at work, the change from a white variety of moth to a black variety of the same moth is an example of species maintenance.

The problem, says Grassé, is that "some contemporary biologists, as soon as they observe a mutation, talk about evolution."[63] This conclusion, says Grassé, "does not agree with the facts. No matter how numerous they may be, mutations do not produce any kind of evolution."[64]

Natural Selection

If Darwin were to be asked what he considered to be his greatest contribution to the field of biology, he would no doubt point with pride to his theory of natural selection. Others, well before Darwin, had speculated on the theory of evolution. Darwin, however, was convinced that he had, for the first time, uncovered the mechanism responsible for the transformation of species. For over a century, the theory of natural selection has been accepted uncritically by fellow biologists and the world at large as the correct explanation for the development of life on earth. So accepting have the scientists been that few bothered to take more than a cursory glance at the assumptions underlying the idea. Had they done so, they might well have been embarrassed by the specious arguments advanced to define the theory. Today, for the first time, scientists are beginning to turn their attention to a critical examination of the concept of natural selection, and their findings are proving quite unsettling to both the theory and the science itself.

To begin with, Gertrude Himmelfarb asks:

> If natural selection is intended to account for the development of species from the simple to the complex and from a low to a higher order of organization, how can it also account for the simple and low? . . . Why have not the superior or higher forms supplanted the inferior or lower?[65]

Himmelfarb uses the example of the hive bee to illustrate her point. Darwin spent a great deal of time praising the hive bee for having developed a nearly perfect instinct. The slow process of natural selection worked to perfect the ability of this little bee to design cells "containing the maximum amount of honey with the minimum expenditure of wax."[66] Darwin marveled at this architectural feat, but was at a loss to explain why other bees, for example the bumblebee, which has not been as good an architect, were still surviving and prospering despite their inferior design skills. Darwin's only rejoinder was that nature left visible traces of its past handiwork on the way to perfecting its forms. But such reasoning doesn't quite square with natural selection, which touts the idea that the more perfect specimen always survives, forcing its less-well-endowed relatives into extinction. Darwin's contention to the contrary, bumblebees, though less well endowed than their hive bee cousins, are nonetheless flourishing, reproducing, spreading their influence, with little apparent concern for their outmoded physiology. Himmelfarb is troubled by the seeming paradox and asks on behalf of the bumblebee and all other creatures throughout the plant and animal kingdom: "Why should there be these living, not dead, remains? Why had not natural selection itself eliminated these imperfect and superseded forms?"[67]

The answer has never been forthcoming, because there are just too many examples of outmoded varieties that have allegedly been surpassed by superior strains thanks to natural selection that have not had the good grace to bow out of the game of life peacefully. It appears, then, that not only the more fit survive, but also the less fit that they replace, which makes for a strange bit of reasoning.

Even more strange is Darwin's contention that natural selection is a gradual process in which each new trait being selected is of some distinct advantage to the individual in its struggle for survival. The problem is that natural selection can never really explain how each little addition that goes into making a complex new part like a limb or organ can by itself be of some advantage. Harvard's Stephen Jay Gould posed the dilemma when he observed, "What good is half a jaw or half a wing?"[68] Certainly, there is no selective advantage whatsoever to partial organs, limbs, or wings, but the theory of natural selection states that every single variation that is selected for must in some way advance the survival capability of the individual. Macbeth points out that Darwin's entire theory hinges on natural selection

> as a mindless process, as the impersonal operation of purely natural forces. If it is mindless, it cannot plan ahead; it cannot make sacrifices now to attain a distant goal, because it has no goals and no mind with which to conceive goals. Therefore every change must be justified by its own immediate advantages, not as leading to some desirable end.[69]

Thus every partial change must in some way be advantageous to the individual and the species. Still, the idea of millions of animals running around with incomplete body parts is enough to turn the head of any right-thinking person, but Darwinists maintain that such was the case in our distant biological past. In fairness to Darwin, it should be noted that he was gravely concerned over just this problem, but kept to his guns despite the fact that he was never able to give a satisfactory explanation of just how natural selection, working gradually, could produce partial parts that were useful to an individual's survival. Nowhere was this dilemma more apparent than in consideration of the eye. Gertrude Himmelfarb goes directly to the heart of the problem.

> Since the eye is obviously of no use at all except in its final, complete form, how could natural selection have functioned in those initial stages of its evolution when the variation had no possible survival value? No single variation, indeed no single part, being of any use without every other, and natural selection presuming no knowledge of the ultimate end or purpose of the organ, the criterion of utility, or survival, would seem to be irrelevant.[70]

The eye is a tremendously complex system, with finely calibrated parts working together with a degree of synchronization that is unmatched by anything that can be artificially produced by human intelligence. Consider the following description by veterinarian R. L. Wysong of what goes into the fashioning of an eye:

> Two bony orbits must be "mutated" to house the globe of the eye. The bone must have appropriate holes (foramina) to allow the appropriate "mutated" blood vessels and nerves to feed the eye. The various layers of the globe, the fibrous capsule, the sclera and chorioid must be formed, along with the inner light sensitive retina layer. The retina, containing the special rod and cone neurons, bipolar neurons, and ganglion neurons, must be appropriately hooked up to the optic nerve which in turn must be appropriately hooked with the mutated sight center in the brain, which in turn must be appropriately hooked up with the grey matter brain stem and spinal cord for conscious awareness and lifesaving reflexes.
>
> Random rearrangements in DNA must also form the lens, vitreous humor, aqueous humor, iris, ciliary body, canal of Schlemm suspensory ligament, cornea, the lacrimal glands and ducts draining to the nose, the rectus and oblique muscles for eye movement, the eyelids, lashes and eyebrows.
>
> All of these newly mutated structures must be perfectly integrated and balanced with all other systems and functioning near perfect before the vision we depend upon would result.[71]

Thus, the eye. Darwin admitted on more than one occasion that he preferred not to have to consider the case of the eye. Writing to Asa Gray in 1860, the English naturalist confided: "The eye to this day gives me a cold shudder."[72] Nonetheless we are supposed to believe that every little mutation along the way to building the complex eye exhibited some useful property that ensured its selection; and that by random chance, and sheer good fortune, all these tiny changes somehow managed to result in the precision and beauty of a fully functional eye, without any preconceived plan and without any knowledge of ultimate purpose. Darwin himself couldn't believe it, even though it was his own theory that advanced the proposition. He wrote:

> To suppose that the eye, with all of its inimitable contrivances for adjusting the focus to different distances, for admitting different

amounts of light, and for the correction of spherical and chromatic aberration, could have been formed by natural selection, seems, I freely confess, absurd in the highest possible degree.[73]

One could draw up a list of tens of thousands of other complex biological systems that utterly defy the idea of gradual development by way of natural selection. In fact, upon close examination, virtually every fully operational system that exists within living things works only as an integrative unit, and the individual parts that make it up appear to exhibit absolutely no value on their own in advancing the survival of the individual or the species.

There are other irresolvable problems that further undermine the validity of natural selection as a mechanism to explain the development of life. For example, natural selection makes no room for long-range considerations. Every new trait has to be immediately useful or it is discarded. The utility of the moment rules over the affairs of life, or at least that's what Darwinists would have us believe. Darwin saw natural selection as extremely economical. It was nature's way of promoting efficiency. Toward this end, it was believed that selection worked to give the survivors just the edge they needed to ensure their perpetuation over their rivals. In other words, just those new traits that were necessary to allow the individual and species to adapt to or cope with the existing environment were selected for. Darwin would have thought it uneconomical and unnatural for a species to exhibit new traits far in advance of any need it might have for them. For Darwin, frugality, not excess, was nature's way. Besides, if an individual were to exhibit traits way in advance of environmental conditions in which they might be useful, then his entire theory of natural selection would be seriously compromised. After all, natural selection was based on the idea of chance combined with pure expediency. There was no room for long-range planning. The contest was always for the moment, the advantage to that creature best adapted to that particular moment.

What happens to Darwin's theory when it is confronted with an organism that is overqualified for its environment? Alfred

Russel Wallace, the co-inventor with Darwin of the theory of evolution, posed just such an example. In comparing the human brain with its alleged ancestor, the gorilla's, Wallace wrote:

> A brain one-half larger than that of the gorilla would ... fully have sufficed for the limited mental development of the savage; and we must therefore admit that the large brain he actually possesses could never have been solely developed by any of those laws of evolution, whose essence is, that they lead to a degree of organization exactly proportionate to the wants of each species, never beyond those wants.... Natural selection could only have endowed savage man with a brain a few degrees superior to that of an ape, whereas he actually possesses one very little inferior to that of a philosopher.[74]

In one fell swoop, Mr. Wallace placed a lethal dagger into his own and Mr. Darwin's theory. Wallace announced that in regard to the human brain, at least, "an instrument has been developed in advance of the needs of its possessor."[75] The venerable Mr. Darwin realized the danger of Wallace's admission and wrote back to him the following message: "I hope you have not murdered too completely your own and my child."[76]

Today a new generation of leading scientists is preparing a long-overdue postmortem on the death Wallace inadvertently inflicted on his own "child." Stephen Jay Gould reflects the thinking of many of his colleagues when he says that it is absurd to continue entertaining the fiction that each and every trait that is carried on through succeeding generations of offspring was somehow selected for its immediate utility. Gould argues that it is time to free ourselves "from the need to interpret all our basic skills as definite adaptations for an explicit purpose."[77]

Competition in the struggle for survival has been considered central to any explanation of the origin and development of species. Darwinists tend to view the world of nature as a battleground where each organism is fighting to maximize its own advantage. Natural selection, they contend, ensures that those

exhibiting the most advantageous traits will survive and produce the most offspring. This rather sophomoric representation
of nature has continued to dominate much of the thinking in science, even though the facts belie the claim. While competition
exists in nature, it is certainly not the only or even the most dominant expression. A century of careful examination of animal interaction has demonstrated a host of behavioral modes.

In their work *Life: Outlines of General Biology,* John Arthur
Thompson and Patrick G. Geddes point out the weakness in the
orthodox view of the struggle for survival in nature.

> What has got into circulation is a caricature of Nature—an exag-
> geration of part of the truth. For while there is in wild Nature
> much stern sifting, great infantile and juvenile mortality, much
> redness of tooth and claw . . . there is much more. In face of limi-
> tations and difficulties, one organism intensifies competition, but
> another increases parental care; one sharpens its weapons, but
> another makes some experiment in mutual aid. . . . The fact is
> that the struggle for existence need not be competitive at all; it is
> illustrated not only by ruthless self-assertiveness, but also by all
> the endeavours of parents for offspring, of mate for mate, of kin
> for kin. The world is not only the abode of the strong, it is also the
> home of the loving.[78]

Natural selection looks good on paper, but as with so many
theories, when exposed to the complex workings of the real
world, the simplicity which made it so convincing in the first
place turns out to be its undoing. For example, proponents of
natural selection would have us believe that there exists some
neat causal relationship between a victim and a predator independent of their surroundings. One can almost visualize the entire contest taking place in an arena fenced off from the vagaries
of the outside world. However, in the real world, the dexterity of
the contestants often has little if anything to do with their sur-
vivability. It makes little difference whether one little ant's legs
are more swift than another's or whether one chimpanzee is
more intelligent than another when a fire or hurricane sweeps
through a forest, killing, indiscriminately, everything in sight.
Natural cataclysms are responsible for a great deal of death and
destruction, but the killing is so random and widespread that it

is just a matter of pure luck which organisms are caught in the path and which are spared. It can hardly be said that those which survived and reproduced were in any sense of the word more fit; they were just more lucky.

At the beginning of any examination of Darwin's theory, the novice is likely to be quite impressed with how well natural selection is able to explain everything. By the end of the examination, however, one is likely to have undergone a 180-degree turn and be of the opinion that natural selection explains almost nothing. Both frames of mind happen to be correct. The fact is, natural selection does indeed explain everything and nothing at the same time. It is a pure tautology. It is instructive to see just exactly how this embarrassing state of affairs came to be.

As the many contradictions to natural selection (mentioned in previous pages) began to surface over the years, its proponents continued to qualify the theory so as to incorporate and defuse the accumulating criticism. Each time natural selection was attacked from a new perspective, its supporters were forced to modify further the basic premise, until all that was left was a mathematical formulation of sorts. Sir Julian Huxley sums up the synthetic or neo-Darwinist conception of natural selection in the following terms: "The struggle for existence merely signifies that a portion of each generation is bound to die before it can reproduce itself."[79]

Not until Nobel prizewinning geneticist T. H. Morgan began to suspect that natural selection was a victim of circular reasoning did anyone in the scientific community even question what was regarded by all as a profound truth. Morgan looked at the definition of natural selection carefully worked out by the neo-Darwinists and then wrote that "it may appear little more than a truism to state that the individuals that are the best adapted to survive have a better chance of surviving than those not so well adapted to survive."[80] Or, as Gertrude Himmelfarb puts it, "The survivors, having survived, are thence judged to be the fittest."[81] Morgan's observation shocked the scientific establishment. It was like proclaiming to the whole world that "the emperor has no clothes." While Morgan helped focus attention on what was regarded as an unthinkable suggestion, a succession of critics have, since that time, taken turns disrobing the proposition, until all that is left of natural selection today is a naked, transparent tautology. C. H. Waddington, one of the greatest biologists of the twentieth century, committed the final act of desacralization when he wrote in a clear and uncompromising fashion the following statement:

> To speak of an animal as "fittest" does not necessarily imply that it is strongest or most healthy, or would win a beauty competition. Essentially it denotes nothing more than leaving most offspring. The general principle of natural selection, in fact, merely amounts to the statement that the individuals which leave most offspring are those which leave most offspring. It is a tautology.[82]

For years, then, scientists had been running around the same circle, faster and faster, until a few, like Morgan and Waddington, began to realize they weren't really going anywhere.

Embryology and Vestigial Organs

Many of the classic arguments that have been used to support evolutionary theory are like malicious gossip. Once in circulation, they feed on themselves. They multiply and expand until they are so pervasive that any attempt to challenge their veracity seems all but futile. Nowhere is this more in evidence than when we examine the field of evolutionary embryology.

"Ontogeny" is the biologist's word for embryological development. "Phylogeny" means evolutionary development. Way back in 1866, Ernst Haeckel, the German biologist and philosopher, combined the two words and proclaimed to the world that "ontogeny recapitulates phylogeny." Translated into simple English, Haeckel's assertion was that during the development of the embryo, it passes through all of the various stages of evolutionary development of its ancestors. According to Haeckel, the embryo represents a moving picture of the entire evolutionary history of life on earth. If one were to watch the human embryo develop, what would pass before the observer's eyes is every single transformation in the long evolutionary sojourn of life, from the emergence of the very first living cell onward. The idea that our total evolutionary inheritance relives itself in the development stages of each and every embryo was appealing, even breathtaking to imagine.

Haeckel's theory took on a life of its own. It was the kind of story that people enjoyed telling whenever the question of evolution arose. In fact, it has remained to this day one of the most popular arguments in defense of evolutionary theory. The idea that "ontogeny recapitulates phylogeny" can still be found in many introductory textbooks in biology, and many a professor still enjoys sharing Haeckel's proposition with his students, thus keeping the story alive long past the time that it was officially abandoned by the architects of biological theory.

Haeckel's argument, which is sometimes referred to as the "biogenetic law," is a myth. There is not a single prominent biologist in the world who is willing to extend so much as a scintilla of credence to it. For over forty years, it has been the subject of ridicule within the scientific establishment; yet it continues to appear in the pages of textbooks and is continually referred to in public discussion and debate. According to Walter J. Bock of the Department of Biological Sciences at Columbia University: "The biogenetic law has become so deeply rooted in biological thought that it cannot be weeded out in spite of its having been demonstrated to be wrong by numerous subsequent scholars."[83]

Whenever the biogenetic law is raised, someone is sure to cite the example of gill slits that supposedly emerge at a certain point in the development of the human embryo as well as in the embryos of other mammals, birds, and reptiles as proof that the embryo is passing through the fish stage on the way to being a bird, reptile, or mammal. It is true that at a certain stage in embryonic development, a series of small grooves, known as pharyngeal pouches, emerge, and that they bear a faint resemblance to similar grooves that appear in the neck area of a fish which later become its gills. But that's where the similarities begin and end. The pharyngeal pouches do not open up into the throat. Instead of developing into slits and gills, they form glands, and also the lower jaw and parts of the inner ear.

While the example of gills is most often used to advance the claim of those who adhere to the biogenetic law, other examples have been trotted forth over the years, and in each case, they have been completely discredited by leading embryologists in the field. Still, acceptance of the law continues unabated as part of the popular understanding of evolutionary theory. Gavin de Beer, former director of the British Museum and one of the world's distinguished embryologists, notes that "until recently the theory of recapitulation still had its ardent supporters . . ."[84] The tenacity with which people cling to such an obvious fallacy led de Beer to conclude that "it is characteristic of a slogan (ontogeny recapitulates phylogeny) that it tends to be accepted uncritically and die hard."[85] A little less charitable is Roy Danson, writing in *New Scientist*. Danson contends that the widespread and persistent acceptance of such a ridiculous conception says as much about the entire field of evolutionary biology as it does about Haeckel's particular contribution. Danson asks, "Can there be any other area of science, for instance, in which a concept as intellectually barren as embryonic recapitulation could be used as evidence for a theory?"[86]

Closely related to the biogenetic law is still another popular myth, the notion of vestigial organs. It is alleged that animals often contain rudimentary organs that are of no useful value to them, and are merely a "vestige" or the remains, if you will, of organs found in some evolutionary ancestor. This rather macabre idea has gained widespread currency over the years. At one time, biologists catalogued what they believed to be over 180 vestigial organs in the human anatomy. Since then, experiments have demonstrated that all of these so-called vestigial organs do indeed perform some necessary function and are in no respect useless vestiges as previously thought. Nonetheless, even today it is not uncommon to hear a proponent of evolutionary theory bring up the appendix or the coccygeal vertebrae as concrete evidence of the existence of vestigial organs.

Despite the fact that the appendix is now believed to perform an important function in fighting infections, this particular organ is still used to justify the idea of vestigial organs. In defense of their claims, adherents of this particular piece of mythology often point out that the human being is able to function as well without an appendix, thus proving its unimportance. As many scholars have observed, human beings can also live without an arm, a leg, an eye, or even a kidney, but that in no way suggests that such things are unimportant or merely vestigial.

Perhaps the most-often-cited example of a vestigial organ is the coccygeal vertebrae, popularly known as the tailbone. In point of fact, it is not a tailbone at all, but for some reason we continue to like to think that this bony little stump at the end of our spine is all that remains of the tail we used to have before we were us. R. L. Wysong points out that, far from being vestigial,

> these vertebrae serve as an important attachment site for the levator ani and coccyques muscles that form the pelvic floor. These muscles have many functions, among which is the ability to support the pelvic organs. Without these muscles (and their sites of attachment) pelvic organs would prolapse, i.e., drop out.[87]

Biogenesis

No discussion of evolution theory would be complete without mention of the famous experiments conducted in the 1950s in which scientists reportedly were successful in creating organic material from inert chemicals. The experiments, performed by Stanley Miller and Harold Urey, have been referred to by evolutionists as proof that life was first generated from nonlife by chance occurrence. The two scientists stimulated a chemical medium containing methane, ammonia, hydrogen, and water with electric sparks producing amino acids and other organic substances. With great fanfare, the world was informed that scientists had finally succeeded in forming life from nonlife, the dream of magicians, sorcerers, and alchemists from time immemorial. Since that historic occasion, virtually every biology student has been made privy to the wondrous secret Miller and Urey had uncovered, a secret that had eluded humanity's grasp over the ages. Great comfort is taken in knowing finally where life originated. In fact, so intent was the need to resolve this question of origins that little effort was extended to probe some of the basic assumptions underlying the Miller/Urey experiment. Had the scientific fraternity bothered to exhibit even a bit of healthy skepticism, they would have seen, at the time, that the Miller/Urey experiment was as much a fictional account of genesis as the long-held myth of spontaneous generation by which scientists of an earlier age had claimed that life arose from dead matter by observing maggots mysteriously appear out of garbage.

It is true that the Miller/Urey experiments were successful in creating organic compounds by running electrical discharges through methane, ammonia, hydrogen, and water. What is not necessarily true is that the conditions they simulated in the chemical medium approximated those on earth at the time life was thought to have emerged, and that the organic compounds they artificially manufactured were the same as those that make up living things.

First of all, there is no way of knowing what the chemical conditions on earth were at the time life first emerged. As John Keosian, professor of Biology at Rutgers, admits, "There is no agreement on what represents primitive earth conditions."[88] On the contrary, there is a great deal of controversy on this score: in fact, so much controversy that, in the final analysis, any experiments designed to duplicate the primeval environment, says biochemist Peter Mora of the National Cancer Institute, are "no more than exercises in organic chemistry."[89]

Miller and Urey's particular exercise in organic chemistry, at first so convincing, has since been subjected to more careful examination and found to be of virtually no value whatsoever in explaining the origins of life.

The first consideration is the chemicals Miller and Urey chose to use to duplicate the chemical conditions on earth at the time life was alleged to have emerged. While the chemical composition on earth in primeval time remains a mystery, it is possible to suggest the likelihood of life emerging under various sets of conditions. In this regard, the particular conditions that Miller and Urey chose to use in their experiment turn out to be totally unconvincing for some very obvious reasons.

To begin with, most scientists would agree that life could not have formed in an oxygen atmosphere. If the chemicals of life are subjected to an oxidizing atmosphere, they will decompose into carbon dioxide, water, and nitrogen. For this reason, it has long been assumed that the first primitive precursors of life must have evolved in a reducing atmosphere, since an oxidizing atmosphere would have been lethal. That is not to suggest that the early earth conditions did indeed preclude oxygen. As Stanley Miller admits, "We do not know that the earth had a reducing atmosphere when it was formed."[90] In fact, it is just as possible that the earth's atmosphere has not changed at all over time. After all, early rock formations contain iron in an oxidized state, suggesting an oxygen atmosphere on the primeval earth. But, in order to posit the theory that life evolved from nonlife, it is essential to assume a reducing atmosphere, because an oxygen atmosphere would destroy the chemicals of life before they could be fashioned into organic compounds by oxidizing or decomposing them back into carbon dioxide, water, nitrogen, and oxygen.

While Miller and Urey's reducing atmosphere overcomes this first giant hurdle, it is immediately faced with a second hurdle, which is insurmountable. Without oxygen, there would be no ozone shield to protect the earth from ultraviolet rays. Without an ozone layer to screen out most of the ultraviolet rays, life could not exist, even on the most primitive level. R. L. Wysong sums up the obvious catch-22:

> If oxygen were in the primitive atmosphere, life could not have arisen because the chemical precursors would have been destroyed through oxidation; if oxygen were not in the primitive atmosphere, then neither would have been ozone, and if ozone were not present to shield the chemical precursors of life from ultraviolet light, life could not have arisen.[91]

To get around this sticky problem, it has been suggested that life might have evolved under water, where it would have been shielded from the lethal ultraviolet rays falling on the earth. Unfortunately, this raises a third obstacle every bit as formidable as the first two. For beginners, the necessary energy catalyst would be absent. Remember, Miller and Urey used electrical discharges to stimulate the chemicals into reacting. They contended that lightning would have performed the same function in the real world. The problem is, lightning will not penetrate through several feet of water, where the ammonia and methane gases are supposed to have taken up residence. Even if it could, which it can't, the chances that any spontaneous biogenesis could take place are virtually zero. In order for life to be generated, the water vapor, ammonia, carbon dioxide, nitrogen, and methane would have to form into amino acids and then combine spontaneously to produce polypeptides. Here's where the problem becomes unsolvable. The "synthesis of polypeptides from amino acids does not take place in the presence of excess water."[92] As organic chemist A. E. Wilder-Smith has pointed out, excess water would serve to break down any polypeptides that might be formed back into the simple amino acids that make it up. Thus, the water prevents the proteins of life from forming.

Wilder-Smith goes on to point out another shortcoming in the Miller/Urey experiment. It turns out that the particular amino acids that Miller formed in his experiment are totally unsuitable for the formation of life. Chemists divide amino acids into levorotary and dextrorotary. The latter are incapable of supporting life. Dextrorotary forms are often lethal. The amino acids of all living forms are levorotary. Wilder-Smith makes the point:

> For biogenesis to take place, *all building blocks* (amino acids) of living protoplasm must be laevorotary.... If even very small amounts of amino acid molecules of the dextrorotary type are present, proteins of a different three dimensional structure are formed, which are unsuitable for life's metabolism.[93]

What this means is that even a combination of levorotary and dextrorotary acids—what chemists call a racemate—would be incapable of synthesizing life. Miller's experiment produced only racemates. In fact, every experiment of a similar kind has produced only racemates; and as Wilder-Smith points out, "Under no circumstances whatsoever is a racemate . . . capable of forming living proteins or life-supporting protoplasm of any sort."[94] It must be emphasized that up to this time it has proven absolutely impossible to form anything other than racemates by stimulating nonliving chemicals with electrical discharges. Harold Urey was asked at a recent conference if he could explain how life could have been formed by the chance combination of chemicals, when all living things require pure levorotary amino acids, whereas in laboratory experiments such as his only racemates are produced by spontaneous processes. His reply is worth repeating: "Well, I have worried about that a great deal and it is a very important question ... and I don't know the answer to it."[95]

G. A. Kerkut sums up the state of the science when it comes to speculation over biogenesis:

> There is, however, little evidence in favour of biogenesis and as yet we have no indication that it can be performed. It is therefore a matter of faith on the part of the biologist that biogenesis did occur and he can choose whatever method of biogenesis happens to suit him personally; the evidence for what did happen is not available.[96]

Miller and Urey's much-fussed-over experiments turn out to be of absolutely no redeeming scientific value when it comes to addressing the question of the origin of life. Like so many other speculations that have characterized the evolutionary literature, their work, if it proves anything, proves how difficult it is to sustain a theory that is confounded at each step of the way by a reality that steadfastly refuses to be accommodated to its governing assumptions.

Mathematical Improbability

Darwin replaced the idea that "this is the best of all possible worlds" with the idea that "all worlds are possible." His theory relied heavily on the notion of probability. He was convinced that given enough time, small changes accumulating over time could account for the transformation of one species into another. But since all these changes in organisms are chance occurrences, without purpose or goal, could one reasonably expect that they could be responsible for the formation of all the highly complex, well-ordered, precisely functioning organisms that make up the plant and animal kingdoms? Darwin staked his professional reputation on it. It was all a matter of probability, he proclaimed. After all, the laws of probability do not preclude any possibility from occurring. Statistically speaking, there is always the chance of something happening, even if it has never happened in the past and despite the apparent unlikelihood of its ever happening in the future. Anyone who has ever considered probability theory knows that if a coin is tossed a million times, the likelihood of getting tails every time, while extremely remote, is, statistically speaking, still a possibility.

The real question, then, is not whether or not evolution is possible but whether or not it is probable. Even considering the fact that the universe is estimated to be 10 billion years old, Sir Fred Hoyle, author of *The Nature of the Universe*, declared that it still does not allow sufficient time for the chance evolution of the nucleic codes for each of the 2,000 genes that regulate the life processes of the more advanced mammals. According to Hoyle, the probability that chance occurrence of random mutations could, through the long process of time, accidentally create the complex ordered relationships expressed through the genetic codes could be likened to the probability that "a tornado sweeping through a junkyard might assemble a Boeing 747."[97]

Still, the Darwinists claim that time is on their side. They point to the fact that the earth is supposed to be nearly 5 billion years old and argue that surely that's enough time for a lot of chance mutations to add up to something of significance. No one would deny that 5 billion years is a long stretch of time, but is it long enough to account for the chance evolution of the entire complex of life in all its myriad forms? The mathematicians would answer with an unequivocal NO. For decades, some of the world's great mathematicians have been pondering and playing with evolutionary claims, attempting to match up time spans with mutation frequencies and the formation of organized living systems, and each time they end by throwing up their hands in utter disbelief. According to all their calculations, the statistical probability that organized life emerged from chance occurrence and accidental arrangement of mutations is virtually zero.

In the world of statistics, events whose probability occurs within the range of $\frac{1}{10}^{30}$ to $\frac{1}{10}^{50}$ are considered impossible. With this as a gauge, let's examine a simple one-celled organism.

> A living cell is a staggeringly complex machine. It consists of thousands of organelles ... and myriads of diverse chemicals all beautifully orchestrated and functioning in a mutually beneficial and orderly fashion.[98]

In terms of information alone, it is estimated that a one-cell bacterium of *E. coli* contains "the equivalent of 100 million pages of *Encyclopedia Britannica.*"[99] A tiny one-cell organism is definitely something to contend with. George Gaylord Simpson tells us that the evolutionary journey leading up to the simplest one-cell organism is as impressive as the rest of the evolutionary trip put together.

> Above the level of the virus, the simplest fully living unit is almost incredibly complex. It has become commonplace to speak of evolution from amoeba to man, as if the amoeba were the simple beginning of the process. On the contrary, if, as must almost necessarily be true, life arose as a simple molecular system, the progression from this state to that of the amoeba is at least as great as from amoeba to man.[100]

Apparently, the mathematical odds more than agree with Simpson's analysis. In fact, according to the odds, the one-cell organism is so complex that the likelihood of its coming together by sheer accident and chance is computed to be around $\frac{1}{10}^{78,436}$.[101] Remember, nonpossibility, according to the statisticians, is found in the range of $\frac{1}{10}^{30}$ to $\frac{1}{10}^{50}$.[102] Needless to say, the odds of a single-cell organism ever occurring by chance mutation are so far out of the ball park as to be unworthy of even being considered on a statistical basis. When one moves from the single-cell organism to higher, even more complex forms of life, the statistical probability shifts from ridiculous to preposterous. Huxley, for example, computed the probability of the emergence of the horse as $10^{3,000,000}$ [103]

Many of the world's great biologists are becoming impatient with the Darwinian notion of evolution by random occurrence and chance mutation. Albert Szent-Gyorgyi, a Nobel prizewinning biochemist, says he can no longer buy the Darwinian interpretation of evolution. Regarding the supposition that random mutations over time do indeed account for the accidental formation of all living things, Szent-Gyorgyi says that he simply cannot accept "the usual answer . . . that there was plenty of time to try everything."[104] This eminent scientist admits: "I could never accept this answer. Random shuttling of bricks will never build a castle or Greek temple, however long the available time."[105]

Several years back, a conference was convened at the Wister Institute of Anatomy and Biology in Philadelphia to address the question of the mathematical probability of evolution theory. In attendance were some of the world's prominent mathematicians and biologists. The latter group was not pleased with what the former group had to say. After making all their computations, the mathematicians concluded that there was not enough time in the entire universe to account for the statistical probability of life forming spontaneously by chance mutation.

As to whether chance mutations, working through natural selection, can, over a sufficient period of time, produce complex living systems, computer scientist Dr. Marcel Schutzenberger of the University of Paris concludes:

> We believe that it is not conceivable. In fact if we try to simulate such a situation by making changes randomly at the typographic level . . . on computer programs we find that we have no chance (i.e. less than $\frac{1}{10}^{1000}$) even to see what the modified program would compute; it just jams.[106]

Using the most advanced computers and the most sophisticated mathematical models, these learned scholars arrived at the following conclusion:

> It is our contention that if "random" is given a serious and crucial interpretation from a probabilistic point of view, the randomness postulate is highly implausible and that an adequate scientific theory of evolution must await the discovery and elucidation of new natural laws.[107]

The findings of the mathematicians were upsetting. After all, evolutionary doctrine owes its very existence to probability theory. For nearly a century, biologists have been preaching that random mutations can account for meaningful structural organization and reorganization over a long enough period of time; and they have used the notion of statistical probability to make their case. Now some of the world's leading mathematicians say there just isn't enough time, statistically speaking, to account for complex sophisticated living systems by the accidental shifting and rearrangement of genetic mutations. Their conclusion serves well as both a summation of and a final epitaph to the neo-Darwinian synthesis.

> Thus to conclude, we believe that there is a considerable gap in the neo-Darwinian theory of evolution, and we believe this gap to be of such a nature that it cannot be bridged within the current conception of biology.[108]

The evidence against the neo-Darwinian synthesis is now so utterly overwhelming that it is astonishing to realize that the theory is still faithfully adhered to and vigorously defended within many sectors of the scientific establishment. Arthur Koestler, one of the distinguished science writers of the twentieth century, speculates as to why the theory continues to linger on. The only plausible conclusion, says Koestler, is that the scientific community would rather continue to believe that "a bad theory is better than no theory."[109] Consequently, says Koestler, "they are unable or unwilling to realize that the citadel they are defending lies in ruins."[110]

In musing as to how a theory as scientifically bankrupt as Darwin's could have ever become the prevailing orthodoxy, the great twentieth-century scientist/philosopher Ludwig von Bertalanffy concluded:

I think the fact that a theory so vague, so insufficiently verifiable and so far from the criteria otherwise applied in "hard" science, has become a dogma, can only be explained on sociological grounds. Society and science have been so steeped in the ideas of mechanism, utilitarianism and the economic concept of free competition, that instead of God, Selection was enthroned as ultimate reality.[111]

PART FIVE

RETHINKING AND REMAKING LIFE

A New "Temporal" Theory of Nature Custom-made for the Biotechnical Age

Our children will not think of the world in a Darwinian way. Nor will they act in the world in a Darwinian manner. For our generation, Darwin served as an all-encompassing explanation of the natural order of things. For our children's generation, it will amount to little more than a historical curiosity. How different the world will seem without a Darwinian screen to shade it. Everywhere our children look they will see the world with a different lens, one tinted to soften the glare of a totally engineered living environment.

Nothing so much distinguishes one generation from another as their differing interpretation of what is meant by the term "alive": its definition, its origin, its purpose, its teleology. For five generations, Darwin's *Origin of Species* has served as the official reference. Whenever our thoughts wandered into the far reaches of the mind's landscape, pondering the ultimate question of what constitutes life, Darwin's cosmology was waiting there with all of the appropriate answers; answers that made sense in a world orchestrated by fire and hissing and belching to the beat of the industrial machine.

Now our children are about to orchestrate a new composition, one played to the whir of a computer console and timed to the tempo of gene synthesizing and cell division. In this new world—this second human epoch—a new interpretation of life is forming. Soon it will stand alone as the sole reference whenever the question arises as to what constitutes the meaning of life.

Darwin's theory of evolution will be remembered in the centuries to come as a cosmological bridge between two world epochs. It represented the culmination of a mode of thought that went hand in hand with the age of pyrotechnology. At the same time, it contained the seeds of the new mode of thinking that will animate the next great world epoch, the age of biotechnology.

When we ask the question What will replace Darwin's theory of evolution? we need only look at what will replace the industrial era in order to find the answer. The age of biotechnology brings with it a new way of organizing nature. And if past history is in any way a guide, this new organizing mode is guaranteed to be sanctified by the construction of a new cosmology that explains the organization of nature in terms that are remarkably congenial with the day-to-day organizing that is going on in that tiny part of the natural world where humanity functions. A new generation, the first of the age of biotechnology, will rest easy, believing that what they are doing to their immediate environment is compatible with the way the whole world has always operated.

In looking at past cosmologies, it is interesting to note that the farther back one goes, the more likely one is to believe that the concept of nature adhered to was a derivative of the cultural setting in which it was formulated. Today some of our leading biologists, as well as historians of science, are even willing to characterize Darwin's original cosmology as more a projection of the cultural and economic biases of the period in which it was written than an independently discovered set of truths about how nature actually works. Still, when it comes to the present, there seems to be an unwillingness on the part of scientists and the public alike to believe that current cosmological reformulations are subject to the same critique. On the contrary, it is argued, and rather pugnaciously, that for the first time in history humankind is exposing the real secrets of nature, once and for all. Borrowing from the world of probability theory, one might argue that while it is indeed statistically possible that such might be the case, it is nonetheless highly improbable.

The more likely scenario is that the new cosmology that will edge out Darwinism and confer grace on the activities and pursuits of the age of biotechnology will be as colored by the technological, cultural, and economic biases of the period as all the cosmologies that preceded it. With this in mind, it should be just as possible to apply foresight as it has been to apply hindsight to the question of cosmological formulation. That is, in assessing the change that bioengineering of life portends for the future of society, what kind of cosmology will be required to legitimize the engineering of all of life by human hands?

What follows is an attempt to lay out the broad elements of the new cosmology that is likely to replace Darwin's theory of evolution. The important point to be made is that the belief that such a conception will emerge is not based on any faith that it represents the universal truths of nature, but only that it reflects the age-old need of every society in every epoch of history to find that unifying principle in nature that will lend an air of unqualified support to its own day-to-day pursuits. That is not to suggest that the new cosmology will not be without any basis of fact in the real world of nature. The problem is that, like all past cosmologies, the emergent theory will claim that the few isolated facts about nature currently being exploited for expedient ends represent the whole of nature's grand operating scheme. From mere snippets, an entire design will be created. The snippets are very real. It is the grand design fashioned from them that is mere deception.

Genotype vs. Phenotype

For as long as the modern synthesis has been around, a tiff of sorts has periodically occurred over the proper relationship between genotype and phenotype. That tiff has suddenly metamorphosed into a full-scale row, dividing experts in the field of biology into two warring factions. It is here that the neo-Darwinist conception is meeting its Waterloo at the hands of a nascent paradigmatic power.

On one side of the battle line are the orthodox neo-Darwinists. They argue that the developing organism (phenotype) is rigidly determined by the genetic information contained within it (genotype). In this traditional view the organism is impervious to all environmental factors and owes its total being to a predetermined genetic blueprint. In other words, the phenotype is merely the product of its genetic information. For the neo-Darwinist, the genotype rules supreme over the process of development.

On the other side of the battle line are the new paradigm shapers, who argue that the genotype is not solely responsible for development of the organism. They are quick to point out that the scientific facts contradict the notion of a simple correlation in which phenotype is a mere expression of genotype. According to the new thinking, while genetic information certainly accounts for a part of the developmental process, it is not the only factor. Many would agree with biologist Søren Løvtrup of the University of Umeå in Sweden, who argues that

> . . . the extranuclear factors are also of ample significance, particularly for the early developmental stages when the body plan is laid down, and it must therefore be concluded that neo-Darwinism cannot possibly claim to account for all aspects of inheritance of evolutionary concern.[1]

The new school of thought contends that "an understanding of evolution requires an understanding of development"[2] and that genetic reductionism is incapable of addressing this larger issue. Their reason for arguing against the idea that everything is reducible to a genetic blueprint is based firmly on the observable evidence.

To begin with, C. H. Waddington points out that "identical phenotypes may have different genotypes . . ."[3] The neo-Darwinists are unable to account for this rather embarrassing fact of life. How is it possible for two organisms to be virtually identical when they contain "wildly different genotypes"?[4] Likewise, how is it possible for two organisms to be very different even if their genotypes are similar? Certainly, if the organism were just a physical copy of its genetic blueprint, these anomalies could not exist. But they do, and because they do, a new generation of scientists are scurrying to find some other, more satisfactory way of explaining the divergence between phenotype and genotype. What they are finding is that an organism's development is influenced by environment as well as by genome. This suggests a revolutionary change in thinking. Remember, according to neo-Darwinist theory, the organism's ability to adapt to the environment is fixed at birth by its genetic makeup. Those organisms fortunate enough to be endowed with a genotype that is better suited to the environment will, by a process of natural selection, live long enough to produce the most offspring, assuring the perpetuation of their genotype in future generations. In the neo-Darwinian paradigm there is no room for ongoing adaptive changes in response to the environment. Just as the Calvinist individual of the Reformation believed in predestination, the question of whether an organism is saved or damned is predetermined, and there is nothing the organism can do to change the outcome during its lifetime.

But everyday observation contradicts this rather facile argument of neo-Darwinism. As biologist M. W. Ho of the Open University in England and mathematician P. T. Saunders of Queen Elizabeth College in London pointed out recently in the *Journal of Theoretical Biology,* organisms are constantly changing their morphological features and behavior in response to the environment. One need go no further than "the muscular development of athletes, and the darkening of skin on exposure to sunlight"[5] for examples of such changes. These types of changes are occurring at every moment of an organism's life and are independent of the genotypical information coded in the organism's cells.

Organisms, then, are not just passive structures, doomed from birth to accept whatever fate their genes have assigned to them. On the contrary, organisms continually modify their behavior and morphology in response to changing environmental stimulus. This line of argument clearly challenges the orthodox view of natural selection. As long as it was believed that the phenotype was merely an expression of the genotype, it could be argued that natural selection was the process of eliminating certain genotypes and preserving others. If we argue that phenotypes are more than a mere expression of their genotypes, and that organisms modify their morphological development and behavior in response to changing environmental stimulus, natural selection becomes much more than a process of choosing among the chance appearance of competing genomes. What is being chosen for also includes the modified behavior and performance of the organism during its development and lifetime. By this new reasoning, Waddington argues that natural selection and evolutionary development is a process that goes beyond simple "chance and necessity" to incorporate "learning and innovation." In other words, what is being selected for is not just chance genetic mutations that happen to confer on some lucky organism the best preadaptation to the environment but also the learned changes over the organism's lifetime that allow it continually to readjust to its environment.

New developments in the field of embryology have done much to dispel the old notion that the organism is "a fixed machine whose structure is entirely predetermined by its genome."[6] In a 1982 cover story in *Newsweek* entitled "How Human Life Begins," the editors made it clear in the opening paragraph that the orthodox view of the gene is being abandoned.

> In a science where the double helix has become the Holy Grail, it may sound like heresy to belittle the mighty gene. But biologists now believe that the laws of development are not as indelibly written into the genes as they once thought: genes are necessary, but they are not sufficient.[7]

The more researchers penetrate the inner workings of embryological development, the more convinced they become that other forces besides genes are at work fashioning the organism. The emerging consensus is that the genes are subordinate in the developmental process and that "the solution to the problem of development lies at the cellular and intercellular level rather than a genetic level."[8] Many scientists even believe that the larger environment in which the mother functions influences the development process within the womb. Neurobiologist Ira Black of Cornell Medical College characterizes these new findings as "stunning from an evolutionary viewpoint."[9]

One of the most intriguing discoveries in embryological research has to do with the relationship of cells to location. In one set of experiments, scientists took a group of cells from the region of the embryo where the neural tube would normally develop and moved them to the region that would normally produce reflex nerves, which make pupils dilate. In their original location the cells would have automatically produced nerves to help control digestion. In their new location they did no such thing. Instead, they produced reflex neurons, thus accommodating the needs of their new site in the developing embryo. In some strange way, the location determined which genes would be turned on within the cells.

In another study, Black found that even the chemicals that flow from the mother to the embryo through the umbilical cord can affect the development of the fetus. When Black placed pregnant rats in a high-stress situation, the "nerve cells in the embryo's gut kept producing one kind of neurotransmitter instead of switching to another as they normally do."[10]

The big question in each of these examples is, What is responsible for turning on one set of genetic instructions instead of another? Remember, each cell in the organism contains all the instructions for duplicating the entire organism. But in the process of development, each of the billions of cells that go into making up the developing organism somehow differentiate into specific functions within the embryo. What force outside the cell triggers the differentiation? Writing in the philosophical journal *Theoria to Theory,* Tony Nuttal of the University of Sussex recounts an experiment with a fruit fly that illustrates the bizarre circumstances surrounding differentiation and the question it poses. Nuttal says that if one takes the eye-producing tissue of a fruit fly embryo and inserts it into the abdominal cavity of an adult fruit fly, it will develop into a winglike structure. Why did a changed location or environment in this and the other examples trigger a change in the genetic readout? According to Nuttal, the only proper conclusion one can come to is that location or environment itself is somehow instrumental in determining how the cells act. In other words, it seems more natural to say that "the pattern imposes itself on the material constituents than to say that the material constituents naturally and fortuitously produce a pattern."[11]

Fields

In the years to come a new term will enter the popular lexicon, taking its place beside the gene as a primary subject for scientific observation. The concept of "fields" was first advanced in biology by the famous Russian scientist Aleksandr Gurvich in 1922. In the sixty years since, it has shown up in scientific literature with increasing frequency.

Fields are a shorthand way of defining the organizational properties of patterns. Fields are invisible, but even though they are nonmaterial, we know that they exist. The notion of fields has been well established in physics ever since Sir Isaac Newton performed his first experiments with gravitation. Although gravity can't be seen, it certainly can be felt, and there is no doubt in any of our minds that its influence is pervasive. Gravity exists everywhere at all times on earth, and it conditions all the organizational patterns of physical phenomena. Even more strange are magnetic fields. Many a child has spent hours on end playing with iron filings and a magnet, transfixed by the invisible force that attracts one to the other.

A familiar experiment illustrates how an invisible field establishes patterns of organization. If iron filings are scattered on a card held over a magnet, they will arrange themselves in a specific pattern along the "lines of force" of the magnetic field. If the filings are thrown away and new filings are scattered on the same card, they will line up in the exact same pattern. Scientists are now beginning to suspect that living cells, like the iron filings, are somehow attracted by an invisible field and arrange themselves in a specific pattern along a line of force established by the field. Donna Jeanne Haraway, author of *Crystals, Fabrics and Fields,* points out that while there is only one kind of electromagnetic field or gravitational field, there are probably "as many potential biological fields as there are organisms and species."[12]

Paul Weiss of Rockefeller University was one of the first to rigorously pursue the idea of biological fields. Trained as an engineer, Weiss brought with him a knowledge of light-wave propagation and electromagnetic field theory. After decades of careful research, Weiss concluded that in regard to living organisms "the patterned structure of the dynamics of the system as a whole coordinates the activities of the constituents."[13] Gone is the old Darwinian notion that the whole organism is merely the aggregrate of each of the separate parts that assemble together into a working machine. In place of this Industrial Age metaphor, Weiss substitutes a novel conception. The parts are not "assembled," they are "integrated." The whole ceases to be an "aggregate," it is rather a "system."

Weiss came to this new understanding by observing the relationship between the separate parts of an organism and the organism's operation as a unit. A simple example demonstrates this relationship.

The next time you take a look at your best friend, consider the rather strange fact that in six months' time not one single molecule that you now see on his or her face will still be there. All of those billions of molecules will have been completely replaced. In the orthodox Darwinian mode of thinking, the face is a solid mass filled in by tiny hard substances called cells that are produced by tiny gene machines tucked inside the nucleus of each cell. The face is thought of as an assemblege of all the specialized products stamped out by each little gene machine. Weiss takes exception to this antiquated view. He asks rhetorically, "How could scalar small-grain units add up all by themselves to the large-grain overall geometric . . . order of the product of their joint construction?"[14] He answers by way of analogy: "It is not unlike asking for the probability of sand grains, free from constraining guidance, arraying themselves perchance in a straight line—and not just once coincidentally, but systematically and consistently."[15] In contrast, Weiss argues that the face is an invisible pattern that somehow coordinates its own parts. How else, he contends, can one account for the fact that the physical substance that makes up the face is continually coming and going but the face itself persists. Weiss believes that, like the iron filings lining up along the field of force of the magnet, the appropriate cells line up in the appropriate places along a biological field and that it is the field that orchestrates the activity within it. Weiss explains how the field operates:

> ... at all times every part "knows" the stations and activities of every other part and "responds" to any excursions and disturbances of the collective equilibrium as if it also "knew" just precisely how best to maintain the integrity of the whole system in concert with the other constituents.[16]

In other words, the individual parts, whether they be molecules, genes, cells, or organs, are continually performing and readjusting their performance to accommodate the larger system which they make up. It is Weiss's contention that the field conditions their performance.

In his book *The Science of Life,* Weiss tells of an experiment in which researchers took a limb bud in the developing embryo and transplanted it to different locations within the sac. In each instance, whether the bud developed into a right or left limb depended "essentially on its orientation relative to the main axes of the body, or more correctly, of the axiate pattern of its immediate surroundings."[17] Other experiments over the years have shown the same results. For example, a severed antenna of a praying mantis can sprout not only another antenna but also a leg instead, depending on the region of the organism to which it is transplanted. Weiss concludes that in the case of the limb bud and the antenna, where two developmental pathways are possible, "which of the two alternatives is to prevail depends on where the cell group stands within the larger complex."[18] In other words, its location within the field determines its behavior.

Weiss is highly critical of the dyed-in-the-wool neo-Darwinists who are still attempting to explain the development and structure of organisms exclusively in terms of genes. Noting that some traditionalists now claim that the stability of the overall pattern of an organism must be due to some hypothetical genes (never yet observed) that perform a "regulatory" and "coordinating" function, Weiss points out the fact that "the cells that carry them [the genes] roll around without fixed bearing."[19] This being the case, Weiss observes:

The patterned form of their ensemble would thus have to arise *de novo* from a mere algebraic shuffle of numbers of gene vehicles— geometry to be derived from algebra—which obviously would presume the intervention of magic—a proposition which a science worthy its name would rather shun.[20]

The only logical alternative explanation, says Weiss, is that the genes and the cells that house them do not organize one another into whole organisms but are rather organized by a biological field that integrates their separate units into a working system.

Another proponent of biological field theory is developmental biologist Brian Goodwin of the University of Sussex in England. Like Weiss he is skeptical of all the claims made on behalf of the genes. Goodwin points out that while genes are the indispensable memory bank for the organism, they provide "no spatial-organizing principles."[21] For that, one must look to biological fields, and Goodwin cites various experiments with the green algae *Acetabularia mediterranea* to make his point. Popularly known as the mermaid cap, this organism is made up of only one rather large cell. At one end of the organism is a branched, root-like structure called the rhizoid; at the other a parasol-like cap. The two ends are joined by a narrow stalk. The single nucleus of the cell is located in the rhizoid. If both the cap and the rhizoid are severed, leaving just the stalk, a new cap is regenerated. Says Goodwin, "The important point is that an undifferentiated stalk can, in the absence of a nucleus [which contains all of the genes], regenerate the highly differentiated cap, so that all the 'decisions' concerning where to synthesize specific enzymes, assemble particular proteins, and regenerate the complex and intricate structure of the cap, are taken in the absence of a nucleus."[22] How could such a thing occur without any genes to coordinate its development? Obviously, says Goodwin, "one must conclude that a morphogenetic field of some kind operates in the cytoplasm of this organism."[23]

Even stranger still, Goodwin has found that biological fields can be affected by electrical potentials and currents. If an electrical differential of 30 mv is produced between the two ends of the stalk, "the cap will form at the end where the external electrical field is more positive."[24] It appears, then, that electrical fields or currents are a feature of biological fields, although Goodwin is quick to suggest that they may not be the only feature. For example, chemical substances have been found to affect developmental pathways and could easily be included along with electrical currents in any description of biological field properties.

Given the mounting experimental evidence, Goodwin and his colleague Gerry Webster, also of the University of Sussex, contend:

> The primary entity which biological science must account for, the living organism with its highly integrated, unified nature and its essential capacity for generation and regeneration, is precisely what these atomic or reductionist ... theories are unable to give any adequate account of.[25]

Writing in the journal *Theoria to Theory*, Goodwin and Webster propose

> that the Darwinian conceptualization of organisms be abandoned and replaced by a structuralist theory of organisms which is based upon the concept of living entities as patterns of relations with specific properties relating to invariance, transformation, and relationship between part and whole. This defines a field theory of organisms.[26]

No one has spent more time on the electrodynamic aspect of biological fields than Yale's Harold Saxton Burr. Like both

Goodwin and Weiss, Burr lambastes orthodox doctrine, which holds that "the chemical elements determine the structure and organization of the organism."[27] Burr argues, like the other field theorists, that such an explanation "fails to explain why a certain structural constancy persists despite continuous metabolism and chemical flux."[28] Burr believes that the place to look for an answer is field theory. His own research and an accumulating body of data coming from other scientists lend support to the thesis that electrodynamic fields play some role in biological fields. In his seminal work, *Blueprint for Immortality*, Burr formulates an electrodynamic field theory for biological development.

> The pattern or organization of any biological system is established by a complex electro-dynamic field. . . . This field is electrical in the physical sense and by its properties relates the entities of the biological system in a characteristic pattern.[29]

According to Burr, "The field is primary and from it stem all the myriad of consequences which are to be seen in Nature."[30] Burr has conducted a number of studies over the years to test the relationship between development of living organisms and electrodynamic fields. In one such study, in cooperation with the Connecticut Agricultural Experiment Station, Burr examined the electrical pattern in seven different strains of corn. Four of the corn seeds were prize strains, three were hybrids. Burr found a "remarkable" correlation between the electrometric activity of the seeds and their growth potential. According to Burr, "The conclusion seems to be inescapable that there is a very close relationship between the genetic constitution and the electrical pattern."[31]

After decades of patient experimental research, Burr is of the opinion that while the chemistry of a living system is "the gasoline that makes the buggy go," it "does not determine the functional properties of a living system any more than changing the gas makes a Rolls-Royce out of a Ford."[32] What, then, determines why a mouse is a mouse and a cat a cat? According to Burr: "The chemistry provides the energy, but the electrical phenomena of the electro-dynamic field determine the direction in which energy flows within the living system."[33]

The thought of invisible fields capable of attracting chemical elements into a unified pattern called a living organism all but boggles the mind. Still, there is a feeling within some sectors of the scientific establishment that Burr, Goodwin, Weiss, Waddington, and the other field-theory proponents are breaking fertile new ground in search of the explanation for the origin and development of species.

Biological Clocks

While field theory is the subject of increasing interest within the biological profession, another area of research has drawn the attention of many scientists. Researchers all over the world are discovering biological clocks, and they are studying them with such relish that even an untrained observer would have to conclude that the experts are on the verge of a major breakthrough in their attempts to shed new light on biological development.

Man has long known that living creatures respond to the rhythmic cycles of nature in a predictable fashion. But it is only recently that scientists have begun to suspect that living organisms contain some kind of mechanism for actually telling time. By the 1950s, enough experimental evidence had been amassed to convince scientists that biological clocks do indeed exist in all life forms from the single-cell organisms to *Homo sapiens.*

In what has since become a landmark set of experiments, biologist Frank A. Brown, Jr., of Northwestern University tested fiddler crabs to ascertain the extent to which they could regulate their activity independent of environmental clues. Brown isolated the crabs from any contact with the sun and moon in order to see how they would react in the absence of outside stimulus. To make it even harder for the crabs, Brown and his colleagues increased the temperature in their isolated environment, knowing that such a rise normally increases the metabolic process. Despite the isolation and the increase in temperature, the crabs maintained their normal rhythmic cycle of twenty-four hours with perfect precision. Brown concluded that "the animals had available some method of time-measuring that was independent of temperature, a phenomenon quite inexplicable in any currently known mechanisms of physiology."[34]

The crabs' performance is by no means unique. After conducting similar experiments with California grunions, Bermuda shrimp, and Australian reef heron, Brown found that they were just as proficient in telling time as the fiddler crabs. Brown suggests that all these organisms, though shut off from the outside world, "had access to outside information as to the time of day (or position of the sun), time of lunar day (or position of the moon), time of lunar month, and even time of year."[35]

In another set of experiments, conducted by psychobiologist C. P. Richter of the Johns Hopkins Medical School, two female squirrel monkeys were tested for the existence of biological clocks. The creatures were blinded at birth, placed in isolation, then monitored for four years. Both monkeys exhibited a natural activity period of 24 hours and 50 minutes and a full-activity cycle of about 30 days without interruption throughout the years of observation. According to Richter, the accumulated data suggested that the internal clocks were in some way related in their evolutionary origin to the lunar cycles. How else could one account for the one-to-one correspondence between the monkeys' daily and monthly cycles and the lunar month, which is 29.5 days? Richter need not have looked so far away from home for a suitable species to test. As it turns out, the average activity period for human beings corresponds to the 24-hour daily cycle, as anyone who has ever had to adjust to jet travel and changing time zones knows all too well. As regards the human menstrual cycle, the average period happens to be exactly 29.5 days, the precise length of the lunar month.

Other experiments have confirmed the notion that living organisms contain their own internal biological clocks, and that while these clocks are set to run in tandem with other periodicities in the environment, they are still able to maintain their precision for a certain length of time even when isolated from those larger rhythms. For example, worms that live in tidal regions will retain their color changes and motion changes, corresponding to the tides, even when transported to a laboratory petri dish, and certain mussels taken from Cape Cod, transported cross-country, and plopped into the Pacific Ocean will continue to correlate their rate of water propulsion with the Atlantic tides for several weeks.

Biologists are beginning to suspect that even cells in the developing embryo have a way of telling time and that their timekeeping ability influences the course of development of the fetus. Many embryologists now believe that cells actually time their movements, and the length of time they spend in one region before moving on to another determines the kind of tissue arrangement they will form. Biologist Lewis Wolpert of the Middlesex Hospital Medical School in London performed an experiment with a chick embryo to prove the point. Wolpert destroyed some of the cells on the embryo's wing tip. The remaining cells continued to divide until they had produced enough progeny to fill in the gap. This task of rejuvenation forced the cells to spend much more time at that particular location than they normally would. According to Wolpert, the long delay fooled the cells into believing they were already at the end of the wing. Apparently, their internal clocks measure out so much time for each particular activity along the line of development. So, by their own calculations they believed they had already arrived at the end of the wing. Consequently, they began to form digits "right on cue." Unfortunately, they were in the wrong place at the right time.

Internal clocks tend to remain set at a certain speed and are generally resistant to small outside influences. Yet it is also true that they are capable of readjusting their timing mechanism in response to radically changing external rhythms. This adjustment is called "entrainment," and it seems that there are a number of environmental stimuli capable of eliciting such changes. Princeton University biologist V. G. Bruce has catalogued some of the more important ones. At the top of the list are changes in light and temperature. Both play a major role in the resetting of internal biological clocks. Researchers have also found that electrical current can affect biological clocks. Experiments on mice and men using electrostatic fields have altered the circadian rhythms in both. Even social interaction among members of the same species have been found to alter the biological clocks of each organism. For example, a spate of recent studies has shown that women living together in close quarters like college dorms tend to menstruate at the same time.

The point is that almost all biological clocks are entrainable within certain limits. It is also becoming increasingly clear that these clocks provide an essential, perhaps *the* essential, mechanism for survival. Speaking of the value of biological clocks to the survival of the organism, Jürgen Aschoff, physiologist and former director of the Max Planck Institute for Behavioral Physiology, contends that "the adaptive significance of circadian rhythmicity is that it enables the organism to master the changing conditions in a temporally programmed world—that is, to do the right thing at the right time."[36]

The old Darwinian notion of survival of the fittest, with its image of brute force and physical stamina fighting it out in a geographically constrained environment, is giving way to a new conception in which survival is determined more by an organism's innate sense of timing. Instead of organisms adjusting to changing spatial contexts, the new view emphasizes the notion of organisms readjusting to changing temporalities. B. C. Goodwin captures the emerging consensus when he asserts that "biological time is inextricably woven into the fabric of biological organization."[37] According to philosopher of time J. T. Fraser, "the capacity to create physiological clocks in response to environmental rhythms appears to be a necessary feature of all life forms."[38]

Finally, it should be noted that while no one has yet demonstrated a conclusive relationship between clocks and fields, many scientists are hinting that the two might be intimately intertwined or even be separate manifestations of some new unifying biological law. Frank A. Brown, Jr., might well have stumbled onto the outlines of an emergent biological paradigm in the concluding paragraphs of his much-praised article "The Rhythmic Nature of Animals and Plants." Brown muses over the relationship between internal biological clocks and the rhythmicity of the environment. Aware that organisms set their own clocks by the rotation of the earth and by the periodicities of the sun and moon, and that they are affected by light, heat, electrical current, and other atmospheric elements, Brown suggests that all these facts together provide "incontrovertible evidence"[39] that all of the rhythms in the universe somehow impress themselves on the organism. The idea that each living thing is somehow affected by "all of the rhythms in the universe" is close to the notion that each organism is somehow affected by fields. The relationship between rhythms (or biological clocks) and fields may be as follows: Rhythms are how fields manifest themselves. Living organisms and all other phenomena, then, can be viewed as particular rhythmic arrangements, which are, in turn, specific manifestations of fields.

A Temporal Theory of Evolution

The neo-Darwinian theory of evolution rests on the assumption that an animal's survival depends on being in the right place at the right time. Unfortunately, the organism is powerless to affect its own destiny. It can do nothing to assure that it will be appropriately outfitted to do battle. According to the neo-Darwinian schema, the organism's structure and behavior are rigidly determined at birth by its genetic program. The organism is plunked down into an environment that is relatively uniform. It lives out its life cycle in rigid conformity with its genetic blueprint, unable to make any substantial innovations in its development or behavior that might allow it to adapt better to changes in its environment. It is a prisoner of its genes. If, however, it is fortunate enough to have been endowed with a genetic program that just happens to fit the environmental circumstances it is forced to contend with, the organism will survive long enough to pass on its genetic endowment to its offspring. Voilà! Natural selection!

The neo-Darwinian conception is rigid and static; yet, ironically, it was Darwin who was credited with breaking the hold of the traditional view of nature as a rigidly determined creation.

Darwin introduced the notion of time into nature. He argued that living species do indeed change over time. Still, he was unable to extend his temporal formulation to include the idea that each organism also changes within its own lifetime. For Darwin, the individual organism was rigidly determined at conception and incapable of changing its predesigned destiny. Yet the species as a whole was continually changing over time as a result of natural selection, sorting, picking, and choosing the most fit among the individual organisms of the population.

Now a new theory of evolution is emerging out of the research findings in embryological development, field theory, and biological clock experimentation. The new theory challenges the notion of fixity anywhere in nature. Where Darwin challenged the idea that the species is fixed and frozen in geological time, the new theory challenges the notion of an individual organism being fixed and frozen in its own lifetime. According to the new theory, nothing is fixed. Every living thing is in flux, continually adjusting to change everywhere else. And unlike Darwin's theory of natural selection, which sees organisms as either adapted to or unadapted to changing physical environments, the new theory sees organisms as either adapting to or failing to adapt to changing temporalities. Darwin's theory of evolution was a spatially conceived cosmology, though it contained a strong dose of temporality. In this sense, it was the heir, albeit a rebellious one, to a long cosmological tradition stretching back to the beginnings of Western civilization. The new theory of evolution, however, is decidedly temporal and marks a profound break in our conception of nature and our place in it.

The transformation from a spatial to a temporal conception of nature marks one of the most spectacular changes in cosmological thinking since the beginning of human history. In order to grasp the magnitude of this change, it is necessary to back up for a moment and review the history of our perception of time.

St. Augustine once wrote, "I know what time is, if no one asks me, but if I try to explain it to one who asks me, I no longer know."[40] For most of human history, time has had to play a catch-up game with space. It has always been relegated to a secondary status, with spatial considerations dominating people's conception of the world. It is not hard to understand why humankind has favored space over time in its conceptual frameworks. As philosopher Milic Capek of Boston University has observed, "Spatial relations seem to us as somehow more fundamental, *more solid,* and easier to grasp than the elusive temporal relations."[41] Because space is easier to grasp than time, the natural inclination of cosmologists has been to place time within space, to give it concrete form, to reduce it to a physical plane. The Eleatic school in ancient Greece was the first to attempt a formal reduction of time to space. According to the Greek philosopher Zeno, time could be properly symbolized as intervals along a spatial segment. Though a convenient shorthand, spatialization of time is an illusory concept born of humanity's need to believe in the lasting nature of things. By reducing time to space, people could relieve themselves of the need to ponder over such things as irreversibility, finiteness, and the notion of death itself. Western man and woman have always viewed space as permanent, so as long as time could be contained within it, humanity could continue the fiction that nothing changes and that everything will continue to exist in a state of everlasting being.

Time is a victim of humanity's need to repress the becoming process. Escape from death has characterized the human sojourn from the very beginning. Time is the ever-present reminder of the futility of our quest for immortality. The passage of time diminishes us; and because, above all else, we seek perpetuity and permanence, the human mind has conceived of myriad ways to downgrade the status of this most fundamental reality.

In some cosmologies, time is viewed as a cyclical process that forever re-creates an eternal image. In others it is viewed as a temporary exile from the lasting state of heavenly beatitude, and in still others as a pale replica or moving image of eternity. Cosmologies have reflected humanity's desire to do away with time altogether. The feat has been accomplished by positing two separate realms of existence, being and becoming, and then claiming the latter to be a part of the former. The philosophy of being is a spatial concept and holds that beneath our experience of time there is an ultimate reality of permanence. The philosophy of becoming is a temporal concept and argues that the ultimate reality of the world is that of pure change. Up until the nineteenth century, virtually every cosmology argued that becoming was part of being or that change was bounded by permanence. Time, then, was always viewed as constrained by space. It could never really go anywhere on its own, being forever confined within a timeless setting. By making time subsidiary to space, humanity could continue to believe that the transitory nature of life was a lesser reality bounded by a larger reality of everlasting permanence.

Since the desire for indestructibility rests at the center of humanity's rejection of time, it stands to reason that the less control humanity is able to exercise over its own future, the less willing it is to recognize the existence of time in its cosmologies. Concomitantly, as humanity has extended its consciousness, enlarged its predictive capacity, and gained greater control over the becoming processes of nature, it has shown a willingness to recognize an increased role for time in its cosmological formulations. In the long history of Western civilization concepts of nature have moved slowly from a fixed spatial orientation to a more fluid, temporal orientation as human beings have gained greater control over their own destiny. As humanity has increased its ability to manipulate time to its advantage, it has elevated the notion of time to a higher and higher status. With greater control over the future, time ceases to be a force to be feared, and comes to be viewed as a tool to manipulate. Time, once seen exclusively as a threat to human permanence, becomes a force people can use to secure their future.

Slowly, very slowly, humanity has been unshackling time from space, giving it greater independence and prominence. The liberation of time from space is reflected in the introduction of a sense of history into natural cosmologies. As people have extended their control over the future, they have increased their awareness of the past. Each succeeding cosmology is more historically oriented than its predecessor. For the Paleolithic hunter, everything was ahistorical and mythical. The Neolithic farmer, and later the Sumerians, began to introduce a touch of history into their cosmologies. The Jews became the first to introduce a cosmology with a history. For the Israelites, creation had a unique beginning point and moved in linear fashion toward a final end, the coming of the Messiah.

It wasn't until Darwin formulated his theory of evolution, however, that time began to loosen its mooring sufficiently to threaten a complete break from its spatial bind. Darwin brought history into nature. For the first time, a cosmology challenged the notion of a fixed creation. The English naturalist argued that living things are subject to history and that they too change over time. Darwin gave concrete expression to a sentiment that was gaining in popularity ever since the philosopher René Descartes proclaimed: "The nature of physical things is much more easily conceived when they are beheld coming gradually into existence, than when they are only considered as produced at once in a finished and perfect state."[42] By introducing the idea that life is the product of historical development and therefore subject to the passage of time, Darwin was raising the specter that time might, in fact, be as important as, or even more important than, space in the conceptualization of nature.

The final liberation of time from space is the result of three related developments over the past one hundred years: the emergence of ecology, the changing conceptions within the field of physics, and the introduction of process philosophy into the thinking of Western civilization. All three have laid the groundwork for a new conception of nature in which time, not space, is the dominant factor.

Ecology, Physics, and Process Philosophy

The term "ecology" made its first appearance in 1866. The brain child of German naturalist Ernst Haeckel, the word is derived from an ancient Greek term that refers to the daily operation of a family household. Haeckel said that, like a household, nature is really a unified economic unit in which each member works in an intimate relationship with everyone else. Haeckel was anxious to delineate ecology as a separate branch of biology whose concern would focus on "the science of the relations of living organisms to the external world, their habitat, customs, energies ... etc."[43] Ecology became the science of living communities or, more specifically, the science of how living communities develop and maintain themselves.

As the pioneers in the new field began to focus their attention on the development of living communities, it became clear to them that time was a key variable. Living systems change over time, and as historian Donald Worster points out, this requires every ecologist to become a natural historian "chronicling the succession of plant societies that occupy a site for a while and then silently disappear into a fossil past."[44]

Where the Darwinians were more concerned with the structure of an organism, the ecologists were more interested in its activity in relation to its environment. In this respect, the ecologists' approach to nature differed substantially from that of other biologists, who much preferred to study animals and plants one by one in isolation. The ecologists were much more concerned with how an organism behaves. To their mind, how an animal or plant looks is much less important than how it acts. Nothing is static, they contended. Everything is in flux, and it is, therefore, the study of movement over time that should concern the natural scientist. Clearly, the ecologists were introducing temporality as a factor to be reckoned with. But it was not until the physicists began to formulate a radical new perspective based on a temporal framework that the biologists began to listen with greater attention to what the ecologists were trying to say.

Classical physics defined matter as impenetrable physical substances. Newton's laws are based on the proposition that two particles can't possibly occupy the same place at the same time because they are each discrete physical entities that take up a certain amount of space. By the early years of this century, the orthodox view of physical phenomena was giving way to an entirely new conception. As the physicists began to probe deeper into the invisible world of the atom, they began to realize that their earlier ideas about solid matter existing in a fixed space were naïve. Much to their surprise, they found that an atom was anything but still. In fact, it became apparent that the atom was not a thing, in the material sense, but rather a set of relationships operating at a certain rhythm. This discovery created quite a stir. Relationships or rhythms cannot exist at an instant. As historian and philosopher R. G. Collingwood of Oxford University has pointed out, relationships and rhythms can exist only in "a tract of time long enough for the rhythm of the movement to establish itself."[45] The Nobel laureate philosopher Henri Bergson once remarked, "A note of music is nothing at an instant."[46] In other words, an individual note cannot by itself be music but requires notes preceding it and following it in time. According to the new physics, this is true for all phenomena. For example, consider the atom. If the atom is a set of relationships operating at a certain rhythm, then "at a certain instant of time the atom does not possess those qualities at all."[47]

With this profound leap in logic, the old idea of structure independent of function is abandoned. The new physics contends that it is impossible to separate what something is from what it does. Everything is pure activity. Nothing is static. Therefore, things no longer exist independent of time but rather through time. The great scientist-philosopher Alfred North Whitehead contrasts the traditional view of physics with the new view that replaced it in the early part of this century.

> The older point of view enables us to abstract from change and to conceive of the full reality of Nature *at an instant,* in abstraction from any temporal duration and characterized ... solely by the instantaneous distribution of matter in space. ... For the modern view, process, activity, and change are the matter of fact. At an instant there is nothing. Each instant is only a way of grouping matters of fact. Thus, since there are no instants, conceived as simple primary entities, there is no Nature at an instant.[48]

According to the new physics, matter is a form of energy and energy is pure activity. Gone forever is the quaint notion of hard substances existing within a "static network of spatial relations."[49] Whitehead delivers the final epitaph to the idea of space as the dominant feature of nature: ". . . the notion of space with its passive, systematic, geometric relationship is entirely inappropriate."[50]

Whitehead summarizes the revolution in thinking that took place in the twentieth century, after thousands of years of slow development, in a few short words: "There is no Nature apart from transition, and there is no transition apart from temporal duration."[51]

It should be emphasized that in acknowledging the overriding importance of temporality, the physicists are in no way suggesting the elimination of space as a meaningful concept. What they are saying is that time and space are abstractions that make sense only in relation to each other. It is impossible, they argue, to explain existence without temporal duration. Time, then, is no longer considered something extraneous to life, but is, rather, an essential attribute of it.

It was a long journey. From time immemorial humanity had believed that time was contained within space. Now Whitehead summarizes the radical new idea that "spatial relations must stretch through time."[52] For the first time in the history of Western civilization, humankind has come to believe that temporal relations are fundamental and that spatial relations are really just convenient mental tools we create to cut into temporality in order to conjure up a fictional moment. "Such instantaneous cuts have their usefulness," says Milic Capek, "but their pragmatic usefulness should not be confused with objective ontological status."[53]

For thousands of years, humankind remained convinced that nature was divided into two categories, structure and function. There was what something was and then there was what it did. Function was seen as an attribute of structure, something derived from it. The new physics reversed the relationship. What something is, physicists proclaim, can be determined only by what it does. Structure does not exist independent of function. It is merely a way of talking about a pattern of activity. The point being made by the new physics is both simple and earth-shattering in import.

Because nature is patterns of activity that exist over time, how it is described by us depends on the length of time we spend observing it. For example, a person walking in the woods is likely to perceive nature very differently from a passenger speeding by in a passing train, or a pilot looking down from a plane racing through the heavens. At greater speeds and at greater distances from the forest, it is possible to get a better view of the outward surface, but at the expense of being able to touch, smell, or hear the many things going on in the interior.

In the old spatial concept of nature, objects were believed to exist independent of time. As a result, it was long assumed that such fixed objects could be analyzed by means of the scientific method and that absolute conclusions could be made on the basis of "objective" analysis. In the new temporal concept, the conclusions one draws about nature are dependent on the relationship between the time frame of the observer and the speed of that which is being observed. Einstein was the first to formalize the idea that what we perceive in the world around us depends on how fast we are going in relation to the things we are observing. In the new temporal frame, "perspective" replaces "objectivity" as the approach to conceptualizing nature. R. G. Collingwood makes the point by way of the following analogy:

> Animals much larger or much smaller than ourselves, whose lives ran in a much slower or a much faster rhythm, would observe processes of a very different kind, and would reach by these observations a very different idea from our own as to what the natural world is like.[54]

Let's take, as an example, a slab of steel. From our perspective it appears to be a hard, firm, static chunk of solid matter. We'd have to stretch our imagination quite a bit to believe that what we are seeing is merely a pattern of activity. Our impression of what we are seeing, however, is related to the temporal frame from which we are observing it. If we could shrink in size to the level of an atom, our microscopic perspective of the steel would be quite different. We would no longer perceive hardness but only the rapid movement of particles all around us.

Anyone who has used a movie projector understands the idea of temporal perspective. If the film is frozen on one frame, you see a picture of a family eating dinner. Everything appears fixed, and it is easy to identify all the structures of the picture—the people, glasses, dishes, and chairs. Speed up the film beyond our normal temporal frame of observation and the images become blurred with people frantically passing one another dishes, putting food into their mouths, getting up and down, exchanging places at the table, going from the dining room into the kitchen, clearing off the table, etc. The activity begins to overwhelm the forms. Eventually the movement is so intense that it is no longer possible to identify specific people, food, or dishes. Everything has fused into a welter of pure motion devoid of all structure.

Let us suppose that time-lapse photography is used to show a plant growing. If the film is run at our normal frame, the plant appears passive and inert. If, however, we speed up the film, the plant will appear quite active indeed, exhibiting constant movement. If we speed up the film even more, the plant will lose its structure entirely and appear to be a whiz of undifferentiated activity zooming across the screen. How we perceive nature, then, is determined by the temporal perspective from which we are observing it.

Advances in the fields of physics and ecology shook the world of philosophy to its very foundations. After centuries of viewing the world from a spatial frame, a new generation of thinkers found itself in the rather awkward position of having to rethink all of its most cherished assumptions. One among their number took on the task of fashioning a new philosophy commensurate with the new temporal frame of reference being advanced in the sciences. His name was Alfred North Whitehead, and history is likely to honor him as the Francis Bacon of the age of biology. Whitehead is the father of process philosophy, an epistemological approach to conceptualizing nature based on the recognition that things no longer exist independent of time, but rather through time.

Whitehead starts with the assumption that all of nature consists of patterns of activity interacting with other patterns of activity. Every organism is a bundle of relationships that somehow maintains itself while interacting with all the other relationships that make up the environment. In interacting with their environment, organisms are continually "taking account" of the many changes going on and continuously changing their own activity to adjust to the cascade of activity around them. This "taking account," according to Whitehead, is the same as "subjective aim." By that Whitehead means that every organism in some way anticipates the future and then chooses one among a number of possible routes to adjust its own behavior to what it expects to encounter. In other words, every organism exhibits some degree of aim or purpose. If an organism were not able to anticipate the future and adjust its behavior to what is about to come, it could not possibly survive all the abrupt changes in the pattern of activity around it.

A proper sense of timing is obviously indispensable to survival and success. All living things are constantly realigning themselves to one another, and if they fail to perceive the shift in patterns, they fall out of step and are left behind in the onward march of life. All of us have known individuals whose sense of timing is dismal. Totally unable to pick up on changing cues in the environment, they are forever out of synchronization with what is going on around them.

Life presents us with an unending stream of novelty. Every passing moment brings with it new conditions that call for some form of assimilation, integration, compromise, or suppression. If the stream of novelty is slow and steady, it is relatively easy to adjust and flow with the current. If the current suddenly turns into a raging rapid, a living organism can be swallowed up before it figures out a way to get on top and ride it out.

To survive, all living things need to be adept at adjusting to novelty. Adjustment requires the ability to "anticipate" what's going to happen before it occurs. This is what Whitehead refers to as "taking account" or "subjective aim." J. T. Fraser, one of the world's authorities on the philosophy of time, notes that anticipating "is evident throughout the animal kingdom."[55] Studies have shown, for example, that rats can anticipate a particular event for up to four minutes and a chimpanzee for up to two days. Like all living things, we humans anticipate in order to bring things under our control. If we can predict the likely outcome of an event before it occurs, we can adjust before the fact, putting ourselves in the best possible position to advance our own interests and survival. Being "one step ahead of the game" was as much a part of the mental makeup of a Paleolithic hunter confronting his prey as it is today for a Wall Street broker confronting his competition. Anticipation allows us to overcome the awful sense of anxiety that comes from not knowing what the next moment or the next day has in store for us. Prediction and foresight are our way of gaining control over the flux of novelty that sweeps over us.

Anticipation works by the process of extension. When faced with an unexpected situation, we do some quick mental reordering. First, we search our repertoire of stored memories for some clues that might be helpful. We attempt to find something in our past experience that might in some way explain what is about to unfold. With the help of our past memory we create an image of what we think is going to transpire. In other words, we attempt to imagine the entire life span of the event: its beginning, its unfolding, and its culmination. By creating a history for something in advance of its materialization, we can pick and choose where and when to intervene in the process.

Control over the ebb and flow of life is the more effective the more one is able to intuit exactly when and where to strike in order to turn things to one's own ends. For example, in the game of football, the key to a good defense rests in the ability of the players to anticipate a play before it unfolds. Even before the ball is snapped, the defensive linemen will assess the way the offense is lining up around the ball. They will then sort through their memory bank in search of past plays that were in any way comparable. They will look for clues right up through the time the ball is snapped in an effort to anticipate the pattern before it unfolds. The more successful they are at drawing a mental picture of the entire history of the pattern from beginning to end, in advance of its execution, the more effective they will be in intervening at the most propitious moment in the process, to turn the play to their advantage.

This constant "anticipation and response" is the central dynamic of all life. A closer look tells us that "subjective aim" is just another expression for mind. Whitehead sees mind (or subjective aim) as existing at every level of life. Organisms are constantly anticipating the future and making choices on how to respond to it. This is mind operating. Whitehead then draws the separate elements of his philosophy into a single vision.

According to Whitehead, mind is not *in* nature, it *is* nature. All living things are patterns of activity. That activity is aimed at anticipating the future in order to adjust to the present, which is just another way of saying that activity is mind at work. When Whitehead characterizes nature as process, he is really characterizing nature as mind at work. Whitehead resolves one of the great paradoxes in Darwinian thinking: How could mere matter produce life and minds?[56] According to Whitehead, mind has been there all along. Nature is pure mind, and each succeeding organism, by dint of its ability to anticipate the future better and adjust accordingly, is exhibiting a pattern of behavior that reflects more and more of the total mind pattern of nature.

Evolution is no longer viewed as a mindless affair; quite the opposite. It is mind enlarging its domain up the chain of species. Where Darwin saw each succeeding species as better able to assimilate a world made up of pure physical resources, process philosophers see each succeeding species as better able to assimilate a world made up of pure mind.

In Whiteheadian thinking, every living thing is a small reflection of the total mind that makes up the universe. It is a pattern woven from the larger pattern. Through subjective aim, each organism seeks to move beyond its own temporal horizon. It reaches out to and is lured by the totality of mind that pervades the universe. Evolution, then, is seen as a movement striving to complete itself. The goal of evolution is the enlargement of mind until it fills the universe and becomes one with it.

Ecology, the new physics, and process philosophy have helped lay the groundwork for the emergence of a new temporal theory of evolution. In his book *Of Time, Passion, and Knowledge,* published in 1975, J. T. Fraser outlines some of the features of this emerging paradigm. His thinking reflects many of the ideas that are just now beginning to converge within the natural sciences.

Fraser begins with the assertion that "life has been characterized from its very beginnings by a striving to increase its control over lengthening periods of future time and thus decrease the uncertainties and attendant tensions of the present."[57] According to Fraser, each species up the evolutionary chain, stretching from the simple protozoa to *Homo sapiens,* is better equipped to anticipate longer futures and a greater multiplicity of novelty in its environment. Under this schema, survival depends on how successful an organism is at narrowing the gap between what it expects and what it eventually encounters in its environment. The better an organism is at anticipating what lies ahead, the better able it will be to adjust its own behavior and ensure its continued survival.

The constant tension between expectation and encounter forces the organism to hone its temporal skills. In other words, its internal clocks are continually adjusting themselves to changing rhythms in the larger environment in order to anticipate future movements that will affect its survival. The more sophisticated its clocks, the greater range and diversity of external rhythms the organism can absorb and the more control it can exercise over its future. In this way of thinking, species are viewed as a hierarchy of increasingly complex temporalities, each better able to anticipate and control its destiny.

Certainly, this seems to make sense. After all, an earthworm is equipped to absorb only a fraction of the periodicities of the environment that a squirrel can absorb. Its biological clocks are rudimentary compared to the latter's and equips it to anticipate only a small range of activity over a narrowly circumscribed horizon. Its ability to control its own future is limited by its ability to anticipate it, and its ability to anticipate it is, in turn, limited by its temporal complexity. The earthworm can pick up light cues in the environment, but the squirrel can pick up spe-

cific images, sounds, and smells as well. This allows it to take in
and contain a far more expansive range of environmental happenings, enabling it to anticipate and control a far greater number of changes in its surroundings.

Different species, then, are to be explained by the difference in
their temporal complexities. But why did one temporal complexity evolve into another, more sophisticated temporal complexity? Fraser offers a new explanation for the origin and
development of species that is very different from the Darwinian
conception. Remember, Darwin argued that organisms are always fighting over limited resources and that evolutionary development results from small variations that favor those organisms
best able to adjust to the constraints imposed by nature. Fraser,
on the other hand, contends that evolution results from organisms adjusting themselves to a scarcity of time, rather than a
scarcity of resources. As to the question of why time becomes an
increasingly scarce commodity, Fraser's contention is that at the
same time that an organism's biological clocks are copying external periodicities, they are also generating some of their own,
thus adding to the number of periodicities in the environment as
a whole. For example, says Fraser:

> Insect mouths, which make the sucking of nectar more efficient,
> influence the nectar balance of the flower; the pheasant's feet, well
> suited to walk on level ground, scratch that ground quite notice-
> ably. . . . Seabirds, which filter out fresh water and excrete con-
> centrated brine, modulate the salt distribution of the ocean, even
> if ever so slightly.[58]

The increase in novelty brought about by the introduction of
new periodicities into the environment increases the tension between expectation and encounter. In other words, the addition
of every new rhythm increases the number of things in the environment that every other organism has to anticipate if it is to
survive. As a result, says Fraser, from the very beginnings of life
until today, "the environmental regularities to be modeled remained forever more numerous and more complicated than the
regularities actually matched."[59] Although he never explicitly

formulates it, Fraser is positing a Malthusian theory of time. The ability of the organism or species to incorporate new periodicities from the environment is always outpaced by the proliferation of new periodicities introduced back into the environment as a result of each temporal adjustment made by every single living organism.

An example illustrates the point. For most of human history population density remained relatively low. People lived in small villages for the most part. By modern standards, the level of social interaction was relatively light. As a result, there was far less novelty to deal with, and people found that they could anticipate what lay ahead with relative ease. In a tranquil country village, the number of future events one has to contend with on any given day is small compared to that faced by an urban dweller in an eight-hour day in midtown Manhattan. As villages gave way to towns and then cities, population density brought with it a great acceleration in social interaction. People found themselves having to anticipate a much wider range of activity. Individual schedules had to greatly accelerate to accommodate the flow of activity around them. However, as each individual accelerated his own routines to keep up with the flow of events, the flow of events accelerated even faster. With the rise of the first modern cities, and population densities in the range of a million or more, time became a scarce commodity as people needed to anticipate and adjust their behavior at a much faster clip. Like the juggler who has to speed up his movements to accommodate each new pin thrown to him, humanity had to speed up its reaction time, as the flow of activity increased, in order to avoid being buried by the deluge.

The point is, as population density increases, the number of potential interactions escalates, requiring each individual to anticipate and adjust to a wider range of possible behavior or rhythms. The individual's own acceleration of activity contributes to the overall intensification of social interaction, requiring everyone else to speed up his own responses. The problem is that the individual cannot possibly absorb the proliferation of new rhythms as fast as they appear. A case in point is life in New York City. The number of new periodicities being added to the environment far exceeds what any individual citizen can assimilate.

Darwin saw evolution as the development of more efficient ways of utilizing limited resources. The new theory views evolution as the development of more efficient ways of utilizing scarce time. Because temporal complexities are continually outpacing temporal adjustments, evolution favors those organisms that are able to respond to changes in the environment in the shortest period of time. Time economy becomes central to a temporal theory of evolution, just as resource economy was to Darwin's theory. Organisms are forced to speed up their reaction time in order to contend with the onslaught of temporal changes in the environment. Organisms continue to reset their biological clocks, and at certain critical thresholds, the improved temporal readjustments result in the transformation to a new, more complex temporality: the evolution of a new species.

A major question posed by the new temporal theory is whether the evolution of a new species is the result of a gradual change in the adjustment of internal biological clocks or the result of a dramatic resetting of internal periodicities. Since the fossil record shows an absence of intermediary forms, it is unlikely that the new theory will settle on the idea that an existing species evolves into a new species by way of a slow, incremental change in its biological rhythms. In order to remain consistent with the abrupt structural changes found in the fossil record, the new temporal cosmology will argue that for transmutation to occur, there needs to be a sudden and dramatic proliferation of external periodicities, forcing existing species to reorganize their entire temporal makeup.

Developments in the field of paleontology over the past decade have laid the foundation for the eventual acceptance of just such a temporal theory of evolution. Uneasy over the fact that the fossil record shows little or no evidence of intermediary forms, a new generation of paleontologists, led by Harvard's Stephen Jay Gould and by Niles Eldredge of the American Museum of Natural History, have advanced the thesis of "punctuated equilibria" as an alternative to the orthodox view of gradual evolutionary development. According to this new theory, species change little if at all over vast stretches of time. Stasis, they argue, is the dominant mode in nature. However, occasionally that stasis is interrupted for a very brief moment of time. Suddenly, a small population from a parent species becomes isolated geographically and begins to evolve rapidly into an entirely new species. This rapid speciation might take place over 50,000 years, which in geologic terms is a mere speck of time. According to Gould and Eldredge, rapid speciation accounts for the absence of intermediary forms in the fossil record. The evolutionary change, they contend, takes place with such lightning speed that the geologic record would hardly have time to record it. Gould and Eldredge's theory, then, rests on the assumption that the history of evolution shows long periods of homeostasis punctuated by short periods of very sudden speciation.[60]

But then the question arises as to what kinds of circumstances could have led to periodic isolation and rapid speciation. The new answer is sudden catastrophe. Unlike Darwin, who argued that the physical environment is relatively constant, a new generation of geologists is arguing that, in fact, the earth's history has been punctuated with severe catastrophic events, some of a global nature. These catastrophes, which probably included massive floods, plagues, earthquakes, volcanic eruptions, meteoric rainstorms, intergalactic disturbances, were likely responsible for the periodic isolation of offspring from parent stock and the rapid speciation that ensued. The idea is that these catastrophic events spawned monstrous genetic mutations within existing species, most of which were lethal. A few of the mutations, however, managed to survive and become the precursors of a new species.

Catastrophe theory and the theory of "punctuated equilibria" are transition arguments on the road to a comprehensive temporal theory of evolution. After all, what are catastrophic occurrences but sudden and dramatic shifts in external periodicities or rhythms? Catastrophes introduce massive, overwhelming novelty into the environment. The sudden speed-up of external periodicities is far greater than what can be anticipated and absorbed by existing organisms. The biological clocks of existing species are disrupted and thrown off track. In the wake of these catastrophic crises, new organisms emerge that are able effectively to incorporate the new external rhythms. These organisms embody entirely new temporalities. They are, in other words, new species.

The idea of catastrophic disturbances stimulating the abrupt appearance of entirely new species fits well with the idea of life as a process of sudden transformation. Past theories of evolution conjure up the notion of gradual, incremental change in morphology taking place over long periods of time. A transformational theory of evolution, on the other hand, implies a dramatic change in a relatively short span of time. Increasingly, scientists will talk in terms of a transformational theory of evolution when examining the history of the origin and development of species.

To sum up: The Darwinian views an organism as a concrete structure that performs a specific function. The newer theory views an organism as a unique complex of behavior patterns. In this newer way of thinking, the behavioral complex that we call an organism is really a bundle of temporal programs that copy some combination of rhythms and periodicities in the larger environment. These temporal programs are predictive devices. They are the organism's way of anticipating the future in order to manage its own survival. Temporal programs are just another way of describing mind—which is to say that mind is what an organism is all about. When we get to the species level, it is argued that each differs from the other in its ability to economize on time or respond faster to a greater range of external periodicities, which is just another way of saying that species differ in intelligence. All species, then, are bundles of knowledge, and each species is distinguished by its intelligence; that is, the speed with which it is able to utilize knowledge to control its own future.

The new temporal theory of evolution is a far cry from the Darwinian conception of living organisms as purposeless, nondirected automata. The temporal theory introduces mind into all living things and declares it to be the moving dynamic of life itself.

Each organism, in the new way of thinking, is a web of temporal programs taken from the entire complex of temporal programs that make up the periodicities of the universe. It is, in short, a specific store of knowledge gleaned from the larger store of knowledge which pervades the cosmos. Nature is no longer seen as the sum total of all of the individual organisms that give rise to it. Instead, each organism is viewed as a pattern of mind woven from the larger pattern of mind that has always existed.

Interestingly enough, the new theories of biogenesis lean toward the idea of the universe as mind. It has become fashionable of late to entertain the rather radical notion that life on earth originated from somewhere else in the cosmic theater. Many of the world's leading scientists, including Sir Francis Crick, the co-discoverer of the double helix of DNA, argue that life migrated to the earth in the form of simple bacteria. Crick believes that the colonization was intentional and directed by some superior intelligence interested in populating distant galaxies. Others agree with Crick that intelligent forms of life existed elsewhere first, but contend that the migration itself was a chance occurrence and suggest that the early precursors to life might have been embedded in meteorites that fell to earth billions of years ago.

In arguing that intelligence existed somewhere else first, and that life here was patterned from it, the scientists are inching closer to the idea of the universe as a mind pulsating with purpose and intention. To be sure, at present they see intelligence as existing in one or more specific locations in the cosmic theater. But just as life on earth is now seen as a pattern woven from a more sophisticated pattern of intelligence that exists at some other fixed location in space, is it not logical to conclude that the intelligence that occasioned our own is, in turn, woven from a still larger pattern of intelligence, and so on? In this way one eventually ends up with the idea of the universe as a mind that oversees, orchestrates, and gives order and structure to all things. If this idea of the universe as mind seems to bear an uncanny resemblance to the idea of fields, it is no accident. When scientists grope to define "fields" in the universe, they are edging closer and closer to the concept of nature as mind. Here is the final juncture where clocks and fields, organisms and environments, fuse together to form the beginning of a new unified theory of biological development, one based on a temporal approach to evolution.

In this new temporal theory, the idea of nature as mind is virtually indistinguishable from the idea of nature as fields. If there is any difference at all between the idea of mind and fields, it probably lies in the fact that a field is a scientific way of framing the idea of mind, making it vulnerable to technological exploitation. By identifying rhythms or periodicities (biological clocks) as the way fields manifest themselves, humankind opens up the possibility of selective intervention and manipulation of the myriad processes of nature on the very deep "field plane."

Cybernetics and the Computer

Ecology, the new physics, and process philosophy all contributed to the liberation of time and helped lay the groundwork for a temporal theory of evolution. Now these sweeping conceptual changes are joining hands with a revolutionary change in the way people go about organizing their day-to-day activity. A great technological transformation is occurring in the world. Humanity is radically changing the way it interacts with the environment. That change is deeply affecting the way people conceptualize the world they live in. In order, then, to fully appreciate the character of the new temporal theory of evolution it is first necessary to understand the technological revolution that is nurturing it.

The Roman statesman Cicero said that by means of our hands we endeavor to create, as it were, a second world within the world of nature. With the steady development of consciousness, humanity increasingly separated itself from its surroundings. From a distance, we began to refashion the various elements of nature into a new construction that bore our own imprint. Our intent was to become invulnerable, and all of Western culture attests to the human desire to overcome the physical limitations imposed by nature and achieve a totally self-contained status. Technology has been the chief means to advance this end.

Technology, quite simply, is tools we design to extend and amplify the human body in order to transform more of nature into ourselves. The clothes we wear and the homes we live in are extensions of the human skin. The containers we use are extensions of our cupped hands. The weapons we use are extensions of our throwing arms.

Technology, then, is a transformer. It allows us to speed up the conversion of natural resources into economic utilities. We develop new and more ingenious ways to transform available matter and energy from the environment to ourselves. Technologies allow us to overcome our biological limits so that we can exercise greater control over the forces of nature. Through technology we conquer time and space. We surround ourselves with technology, and it, in turn, surrounds us with a remade second nature. It's no wonder humanity has always been so admiring of its tools. We gaze upon them with a mixture of awe and respect, as we would our own body organs. They are our enlarged body, and through them we are able to inflate our sense of ourselves well beyond the proportions nature endowed us with.

We watch over our tools, constantly seeking to hone them to new levels of performance. They become our obsession as we become the increasing object of our own attention. As we outfit ourselves with more sophisticated technological garb, nature seems to shrink in size and our own gait seems to stretch out over larger terrain.

Technology translates our innermost desires for plenitude, security, and self-perpetuation into concrete forms that we can take hold of and believe in. The more successful we are at trapping, converting, and assimilating the world around us, the more assured we are of continued abundance, a safe future, and a self-perpetuating existence. Technology is our way of proving to ourselves that we will prevail. As such, it has always enjoyed a most exalted status.

Technology helps define the world around us. That's because our understanding of the environment is deeply influenced by the way we go about organizing it. Abraham Maslow once remarked, "To him who has only a hammer, the whole world looks like a nail."[61] It is not surprising, then, that as we have changed our organizing tools, our cosmologies have changed as well, reflecting the new approach we are using to capture and transform parts of nature into ourselves. Through our technology we project ourselves into nature. Through our cosmology, we turn our technological relationship with nature into timeless truths. Our ideas about how the universe operates are conditioned by the way we are operating in a tiny fraction of the universe at any given moment. We convince ourselves that the way we come to fashion nature conforms to the way nature itself is fashioned.

The new temporal theory of evolution, like other concepts of nature that have preceded it, expresses the new organizing relationship we are establishing with our environment. We are undergoing a revolutionary transformation in our mode of technology, and that organizational change is laying the base for the age of biotechnology and the new temporal cosmology that will accompany it.

Two historic developments took place in the early years of the 1950s that together marked the beginnings of the age of biotechnology. In England, two young scientists, James Watson and Francis Crick, discovered the double helix, fully exposing the gene to human scrutiny for the first time in history. Across the Atlantic, the first working computer was installed in Blue Bell, Pennsylvania, for use by the U.S. Census Bureau. For the next three decades, developments in biology and the computer sciences traveled along separate lanes of the same developmental pathway. Today, biology and the computer sciences are about to merge into a single dynamic. The fusion of the computer and living tissue signals the end of the age of pyrotechnology and the beginning of the age of biotechnology.

The first Univac computer installed at Blue Bell, Pennsylvania, was a gangly affair. It was eight feet high, seven and one-half feet wide, and fourteen and one-half feet long. A journalist covering the unveiling said that it looked something like "a combination pipe organ console, a Linotype machine and a telephone switch-board."[62] The computer remained a curiosity until the advent of the transistor and the microchip. By 1959, 6,000 computers were on line; by 1966, more than 15,000; by 1970, over 80,000, per-forming tasks in more than 3,000 different categories. Today, there are millions upon millions of computers pervading every facet of life. IBM announced that it had more orders on hand for 1980 models than the total amount of all of its deliveries from the years 1950 to 1979.

The 1980s saw the spread of the electronic computer from of-fice to home. There are presently over a hundred companies manufacturing home computers, and as one retailer remarked, "Some day soon every home will have a computer. It will be as standard as a toilet."[63] Futurist Alvin Toffler predicts that within the next few years computers will be built into

> everything from air-conditioners and autos to sewing machines and scales. They will monitor and minimize the waste of energy in the home. They will adjust the amount of detergent and the water temperature for each washing machine load. They will fine-tune the car's fuel system. They will flag us when something needs re-pair. They will flick on the clock radio, the toaster, the coffee maker and the shower for us in the morning. They will warm the garage, lock the doors and perform a vertiginous variety of other humble and not-so-humble tasks.[64]

Within the next two decades the electronic computer will connect factory to office to store to home in one interrelated complex. The computer will be found everywhere that requires any kind of decision-making. The computer is an artificial ex-tension of the human mind that is about to be tucked into every nook and cranny of the world around us.

The human environment has become increasingly complex, requiring greater anticipation and faster decision-making. Since the human mind can't possibly be everywhere at the same time, the electronic computer will be increasingly relied on to manage our future. The impact of the computer revolution is compara-ble to the introduction of the steam engine, which launched the

Industrial Revolution. With the steam engine humanity replaced muscle power with an inanimate source of energy.
Today, with the computer, humanity complements the mind
with an artificial intelligence.

The electronic computer marks a new chapter in human history. For the first time we have developed a "technological"
means of projecting the human mind directly into nature. This
new approach to organizing the environment is radically changing our conception of nature. Remember, the industrial era was
characterized by the technological projection of the human
body into nature. When the machine replaced muscle power,
our cosmology changed in turn, reflecting the new method of organization. Darwin constructed nature in the image of the industrial machine. The new temporal theory of evolution is
reconstructing nature in the image of the electronic computer.
Nature as "matter in motion" is being replaced by nature as
"mind in action."

The change in technological modes is already effecting a radical change in work-place psychology. "Programming a computer" requires a far different psychological orientation than
"attending a machine." In the industrial era, workers were
rewarded for their diligence and reliability. "Industrial" comes
from the Latin word *"industrialis."* Today, "industriousness" is
no longer held in as high esteem as it was in our parents' and
grandparents' generation. In fact, when we refer to someone
today as industrious, we often think of him as a hardworking,
plodding type to be respected but not venerated. Industrious behavior is behavior generally associated with physical labor. In an
age when the human body was projected onto the environment
in the form of "the machine," it's not hard to understand how a
premium would be placed on being industrious.

The psychological motivation that goes with the computerized society is of a very different sort. The new worker strives
to be creative, resourceful, integrative, and informed. The new
psychological orientation values mental acumen over physical
strength and fits nicely into an age in which the human mind is
projected onto the environment in the form of "the computer."

In psychological terms, we are moving from the "industrious" age to the "informed" age.

The psychological reorientation of the worker is just a small part of the sweeping changes that are taking place as the human family adjusts to the shift from a society organized by the industrial machine to a society organized by the computer.

While the electronic computer is fast replacing the industrial machine as the critical operating technology of civilization and is forcing a basic change in the psychological orientation of work-related activity, it is also becoming the chief metaphor for the reconceptualization of the origin and development of species. It is no mere coincidence that many of the operating principles that animate the computer happen to be the same operating principles that biologists now claim are the basis of all living systems. The cosmologists are once again borrowing the organizing technology of the society and "projecting" it onto nature. To the question How does nature operate? the new answer is that it operates in a manner similar to the electronic computer.

In order to fully digest the extent of the projection, it is necessary to delve into the operating principles that underlie the computer revolution. Those principles first took concrete form during World War II, when teams of engineers and scientists were assembled by the government with a mandate to devise new ways of organizing an increasing array of disparate information into an intelligent, efficient mode of operation. The undertaking was called "operations research," and from it a new approach to organization emerged; it was called cybernetics, and it provided the operating principles for the computer revolution.

"Cybernetics" comes from the Greek word *"kybernetes,"* which means "steersman." It is a general theory that attempts to explain how phenomena maintain themselves over time. According to philosopher Carl Mitcham of St. Catherine's College, cybernetics is not concerned with "what a thing is but how it behaves."[65] Cybernetics reduces behavior to two essential ingredients, information and feedback, and claims that all processes can be understood as amplifications and complexifications of both.

M.I.T. mathematician Norbert Wiener, the man who popularized cybernetic theory, defined information as the

> name for the content of what is exchanged with the outer world as we adjust to it, and make our adjustment felt upon it. The process of receiving and of using information is the process of our adjusting to the contingencies of the outer environment, and of our living effectively within that environment.[66]

Information, then, consists of the countless messages that go back and forth between things and their environment. Cybernetics, in turn, is the theory of the way those messages or pieces of information interact with one another to produce predictable forms of behavior.

According to cybernetics theory, the "steering" mechanism that regulates all behavior is feedback. Anyone who has ever adjusted a thermostat is familiar with how feedback works. The thermostat regulates the room temperature by monitoring the change in temperature in the room. If the room cools off and the temperature dips below the mark set on the dial, the thermostat kicks on the furnace, and the furnace remains on until the room temperature coincides once again with the temperature set on the dial. Then the thermostat kicks off the furnace, until the room temperature drops again, requiring additional heat. This is an example of negative feedback. All systems maintain themselves by the use of negative feedback. Its opposite, positive feedback, produces results of a very different kind. In positive feedback, a change in activity feeds on itself, reinforcing and intensifying the process, rather than readjusting and dampening it. For example, a sore throat causes a person to cough, and the coughing, in turn, exacerbates the sore throat.

Cybernetics is primarily concerned with negative feedback. Wiener points out that "for any machine subject to a varied external environment to act effectively it is necessary that information concerning the results of its own action be furnished to it as part of the information on which it must continue to act."[67]

Feedback provides information to the machine on its actual performance, which is then measured against the expected performance. The information allows the machine to adjust its activity accordingly, in order to close the gap between what is expected of it and how it in fact behaves. Cybernetics is the theory of how machines self-regulate themselves in changing environments. More than that, cybernetics is the theory that explains purposeful behavior in machines.

It was in a landmark article published in the *Philosophy of Science* in 1943 that Wiener first introduced the notion that machines can exhibit purposefulness. Wiener defined purposeful behavior as "a final condition in which the behaving object reaches a definite correlation in time or space with respect to another object or event."[68] For Wiener, all purposeful behavior reduces itself to "information processing."

> It becomes plausible that information ... belongs among the great concepts of science such as matter, energy and electric charge. Our adjustment to the world around us depends upon the informational windows that our senses provide.[69]

After careful deliberation, Wiener concluded that "society can only be understood through a study of the messages and the communications facilities which belong to it."[70] It's no wonder he came to view cybernetics as both a unifying theory and a methodological tool for reorganizing the entire world. Apparently, the succeeding generation of engineers and scientists fully concurred. With the aid of the computer, cybernetics has become the primary methodological approach for organizing economic and social activity. Virtually every activity of importance in today's society is being brought under the control of cybernetic principles. "Information processing" via the computer is fast becoming the hallmark of our technological society. Nowhere is this more in evidence than in the economic system. Once considered an adjunct to the management of large-scale economic organizations, information processing has risen to the top of the corporate pyramid and now defines the organization itself. Corporations are increasingly viewed as information systems. In her book *Systems Analysis in Public Policy* sociologist Ida Hoos of the University of California observes that "the management of information has become equated with and is tantamount to the management of the enterprise."[71] According to Hoos, "management *of* information has receded from view and management *by* information has become the mode."[72]

Cybernetics has not only changed the way we go about organizing the world, but has also affected the way we go about conceptualizing it. To begin with, the operating assumptions of cybernetics are antithetical to the orthodox view of the relationship between parts and whole. During the industrial era, it was assumed that the whole was merely an aggregate of the assembled parts that made it up. Cybernetics, in contrast, views the whole as an integrated system. The constant feedback of new information from the environment and the continual readjustment of the system to the environment set up a circular process in contradistinction to the linear mode of organization that characterized the Industrial Age. The self-correcting circularity of this new mode of organization "blurs the distinction between cause and effect."[73] According to the cyberneticians, in an increasingly complex environment it is no longer possible to entertain the simple fiction that one event in isolation leads to another event in isolation. We are now coming to realize that every event in some way affects everything else. Because everything is interrelated, it is necessary to organize activity into integrated systems.

Cybernetic thinking has made its way into the very construction of machines. In the industrial era, machines were made up of many individual parts assembled into a working whole. Today, cybernetics has combined with computer design and electricity to "integrate more and more functions into fewer and fewer parts, substituting 'wholes' for many discrete components."[74] The watch is a prime example of the transition in thinking and design that has occurred. Alvin Toffler points out that "whereas watches once had hundreds of moving parts, we are now able to make solid-state watches that are more accurate and reliable—with no moving parts at all."[75]

In an article entitled "The Development of Cybernetics,"

Charles R. Dechert, professor of Political Philosophy at Catholic University of America, sums up the importance of the new set of organizing principles that have replaced the "assembly line" mentality of the Industrial Age: "Cybernetics extends the circle of processes which can be controlled—this is its special property and merit."[76] Lest there be any wavering doubts as to the efficacy of this new organizational form, futurist Robert Theobald cautions us that there is no way to turn back the clock. "Let us be very clear: the only way to run the complex society of the second half of the twentieth century is to use the computer."[77] Increased reliance on the electronic computer ensures the institutionalization of cybernetic principles as the central organizing mode of the future.

It's worth noting, at this point, that without electricity, cybernetics and the computer revolution would have been impossible. Electricity, observed Marshall McLuhan, provided the means for the "instant synchronization of numerous operations," and by so doing, it "ended the old mechanical pattern of setting up operations in lineal sequence."[78]

Electricity greatly advances humanity's technological projection of itself. For tens of thousands of years humans had pursued invention in order to amplify and extend their bodily limbs. With electricity humanity began an extension of a different kind. Marshall McLuhan points out that "with the arrival of electric technology, man extended, or set outside himself, a live model of the central nervous system itself."[79] As an extension of the nervous system, electricity allows us to greatly enhance our capacity to anticipate and adjust to increasing levels of activity in the surrounding environment.

All earlier modes of technology were necessarily "partial and fragmentary" because they were unable to overcome their spatial context. Because electricity moves at the speed of light it is able to transcend spatial context. Electricity allows humanity to skip over long stretches of space virtually instantaneously. Marshall McLuhan was one of the first to observe that with electric media humanity is able for the first time to "abolish the spatial dimension."[80] Today a person in the United States can turn on his TV set and watch an event taking place 8,000 miles away in

Africa via electrical impulses transmitted by satellites. As a result, says McLuhan, "what emerges is a total field of inclusive awareness."[81] By subsuming space, electricity redefines all activity as pure process.

Electricity, cybernetics, and the computer translate into the organizing energy, the organizing principles, and the organizing mechanism for the new age of civilization.

When Norbert Wiener published the first edition of his book *Cybernetics,* he included a subtitle: "Control and Communication in the Animal and the Machine." Wiener was convinced that the operating principles of cybernetics could be successfully extended from the engineering field to the life sciences. His goal was to reformulate biology in engineering terms, making it subject to rigorous mathematical analysis. Wiener once remarked that the only real difference between fire control in antiaircraft gunnery and biological processes was the degree of complexity governing their respective information-sorting and feedback capacities.

Wiener dreamed of unifying engineering and biology, and apparently many technicians in both fields were anxious to share his vision. In the thirty-five years that have elapsed since he sketched out his grand design, biological thinking has been recast in the image of engineering technology. In their book *Current Problems in Animal Behavior* zoologist William H. Thorpe and psychologist Oliver L. Zangwill of Cambridge University reassess the impact of engineering on the field of biology and conclude that the life sciences have all but succumbed to the operating assumptions of the technologists. The two scholars note that "principles derived from control and communications engineering are being increasingly brought to bear upon biological problems and 'models' derived from these principles are proving fertile in the explanation of behaviour."[82] According to Thorpe and Zangwill, scientists in both fields are finding common ground "under Norbert Wiener's banner of Cybernetics."[83] As a result, the distinction between engineering principles and biological principles is beginning to blur.

The fact of the matter is, biology is being totally revamped along engineering lines. Philosopher of science Marjorie Grene of Harvard University suggests that as the engineers improve the capacity of machines to regulate their own performances, they become more and more convinced that their own handiwork behaves much like living systems. Ergo, they conclude that the same operational guidelines they are imposing upon their technology must bear some correlation with the operating guidelines that animate living systems. According to Grene, a new way of thinking has permeated the biological field.

> It says in effect: look to engineering, to blueprints and operational principles . . . for the sources of your theoretical models in biology, much as Darwin drew on the works of sheep breeders and pigeon fanciers as a source for Natural Selection.[84]

Perhaps the best way to express the extent to which engineering has been able to recast the field of biology in its own image is to take a look at the word "performance." Engineers use this word to refer to the activity of machines. Biologists in contrast have traditionally relied on the word "behavior" when referring to the activity of living organisms. Performance conjures up the idea of purposeful activity designed to meet a specific objective. Behavior, on the other hand, often connotes the image of undirected activity without specific goals. Experimental psychologist R. L. Gregory of Cambridge University notes that biologists are increasingly using the two words interchangeably, "the reason almost certainly being the influence of cybernetic ideas, which have unified certain aspects of biology and engineering."[85] The two scholars go on to say that the term "performance" is being relied on increasingly as biologists begin to redefine living organisms in terms of relative efficiencies. Clearly the engineering mentality has taken hold within biology; in fact, so much so that, as Thorpe and Zangwill point out, living organisms are more and more being described in terms of their thermal efficiency, information efficiency, capital costs, running cost, and other technologically conceived criteria.

It is a measure of the immense influence that the engineering field has had on biology that most biologists have come to accept cybernetic principles as an operating language for their discipline. Cybernetics is providing a brand-new form of communication for biologists, and it is the shared acceptance of this new language that is laying the groundwork for the reconceptualization of nature and the acceptance of a temporal theory of evolution.

Biologists now view living organisms as information systems. W. H. Thorpe defines living organisms as things that ". . . absorb and store information, change their behavior as a result of that information, and . . . have special organs for detecting, sorting and organising this information . . ."[86] Biologists have been completely won over to the idea that all phenomena are reducible to information processing. The older Newtonian model, which viewed nature as "the movement of a particle under the action of a force," has been replaced with a new model that defines nature as "the storage . . . and the transmission of information within a system."[87] When one stops to consider that information is a nonmaterial thing, the full impact of the revolution in thinking begins to come into focus. Because it is nonmaterial, information does not exist in a static spatial context in the sense that Newton had in mind when he defined the world in terms of matter in motion. When a biologist talks about living organisms as information systems, he is saying that they are instructions or programs that "describe a process and, further, instruct that this process should be done."[88] When a biologist talks about process, he is referring to something that takes place over a period of time. Therefore, living systems in the new way of thinking are information programs that unfold in a predictable manner over time.

"The most important biological discovery of recent years," says W. H. Thorpe, "is the discovery that the processes of life are directed by programmes . . . [and] that life is not merely programmed activity but self-programmed activity."[89] If the word "life" were removed from the above quote, one might well suspect that what Thorpe and the other biologists are talking about are computers with their information processing, their programs, their self-regulating activity. Indeed, the computer is what they are talking about in a very roundabout way. It has become the new metaphor for defining life and will be every bit as convincing to a generation raised on videogames and pocket calculators as the industrial machine was convincing as a metaphor for defining life among those conditioned by the industrial era. In Darwin's day, life was viewed as an aggregate of separate, interchangeable parts assembled into a working whole. Today, life is viewed as a code containing millions of bits of information capable of being programmed in a number of specific ways. We are experiencing the transformation from industrial machine to computer, from assembling to processing, from space to time, and from the cosmology of the Industrial Age to the cosmology of the Age of Biotechnology.

The battle over competing biological paradigms is as much a struggle over competing languages and metaphors as anything else. The neo-Darwinists continue to use the language and metaphors of the Industrial Age, while a new generation of scientists are using the language and metaphors of the Age of Biotechnology. It's a battle between those who continue to think in terms of "the best-built machines" and those who think in terms of "the best-designed programs." Philosopher Kenneth M. Sayre of Notre Dame University is from the new school. He says, without qualification: "the fundamental category of life is information. . . ."[90] Other scientists of his ilk are anxious to prove the point and are setting out to redraw the map of life. Any self-respecting computer programmer would be tickled with the picture that is emerging.

The French biologist Pierre Grassé has laid out a detailed presentation of the new approach to the conceptualization of nature, using the language of cybernetics. Grassé begins by framing all of life in cybernetic terms: "Information forms and animates the living organism. Evolution is, in the end, the process by which the creature modifies its information and acquires other information."[91]

Grassé then goes on to develop a cybernetic model of a living organism. He starts with the strands of DNA that make up the genetic code. According to Grassé, the code represents the intelligence of the species. Grassé is willing to concede that DNA is

". . . the depository and distributor of the information . . . ," but he takes exception to James Watson, Francis Crick, and many of the surviving neo-Darwinists who contend that it is also the "sole creator."[92] Grassé compares the genetic code of an organism to a library and argues that neither one fabricates information; they are merely repositories of information received from the outside. Both DNA and the library classify and store. It is at this point that Grassé applies the principle of "feedback" to living systems.

> DNA has to receive messages either from other parts of the cell or from organs . . . or from the outside world (sense stimuli, phero-mones, etc.). Of itself, by what miracle could it generate informa-tion adequate to performance of a given function?[93]

To hammer his point home, Grassé feels compelled to use the computer as an appropriate reference point.

> The computer is limited in its operations by the program control-ling it and the units of information fed into it. To enlarge its possi-bilities, its contents have to be enriched. What is new comes from outside.[94]

Grassé concludes that the living organism, like the computer, has "to be programmed and fed with external information in order for novelties to emerge."[95] The picture he sketches is a cy-bernetic model of life; a circular process in which the genes, the organism, and the environment continually feed information back and forth, allowing the organism to regulate itself in re-sponse to changing external cues. After cataloguing all the ways that external factors influence the genes and vice versa, Grassé acknowledges the role that cybernetics has played in helping to redefine the very operational design of life itself. He concludes that "the cybernetic model, of which philosophy has not yet fully taken advantage, is applicable to all kinds of biological sys-tems . . ."[96]

The new cybernetic model of living organisms is the opera-tional counterpart of Whitehead's notion of "subjective aim." Cybernetics secularizes Whitehead's process philosophy by turning a metaphysical insight into a technological modality. Remember, Whitehead's contention is that all living things are constantly "taking account" of the many changes going on and continually adjusting their own performance to anticipate future states. According to Whitehead, this "process" is at the center of all activity in nature and is really a shorthand description of "mind" at work. Cybernetics reduces Whitehead's description of mind in nature to quantifiable proportions, replacing any vitalistic or spiritual embodiment with a purely technological definition of behavior.

By reducing all activity to information feedback and processing, cyberneticians are saying that the most important defining characteristic of mind is the ability to anticipate and respond to changing conditions over time. For the cybernetician, information feedback and information processing serve as a kind of all-embracing technological description of how the mind operates at every level of existence. Information feedback and processing, in turn, are ways of describing how an organism changes over time. In fact, if we dissect the word "information," we find that it means "in" "formation." The cybernetician views living organisms as "in" "formation." A living organism is no longer seen as a permanent form but rather as a network of activity. With this new definition of life, the philosophy of becoming supersedes the philosophy of being, and life and mind become intricately bound to the notion of change over time.

While cybernetics is largely concerned with how systems maintain themselves over time, it also makes room for the idea of evolutionary change in systems by way of positive feedback. Ilya Prigogine, a Belgian physical chemist, has devised a theory of dissipative structures to explain how cybernetic principles can incorporate the notion of evolution as well as homeostasis. According to Prigogine, all living things and many nonliving things are dissipative structures. That is, they maintain their structure by the continual flow of energy through their system. That flow of energy keeps the system in a constant state of flux. For the most part, the fluctuations are small and can be easily adjusted to by way of negative feedback. However, occasionally the fluctuations may become so great that the system is unable to adjust and positive feedback takes over. The fluctuations feed off themselves, and the amplification can easily overwhelm the entire system. When that happens, the system either collapses or reorganizes itself. If it is able to reorganize itself, the new dissipative structure will always exhibit a higher order of complexity, integration, and a greater energy flow through than its predecessor. Each successive reordering, because it is more complex than the one preceding it, is even more vulnerable to fluctuations and reordering. Thus, increased complexity creates the condition for evolutionary development.

Like Prigogine, many scientists are coming to view evolution as the tendency of all living systems to advance toward "increased complexity of organization."[97] Organizational complexity, in turn, "is equivalent to the accumulation of information . . ."[98] In other words, evolution is seen as improvement in information processing. The more successful a species is at processing more complex, more diverse kinds of information, the better able it is to adjust to a greater array of environmental changes. By this new way of thinking, the key to evolution itself is to be found in how information is processed. Negative feedback leads to stasis. Positive feedback leads to transformation. The upshot of this reformulation, says John Ford, former executive director of the American Society for Cybernetics, is that "together with philosophy cybernetics [becomes] . . . the basis of the evolving theory of development."[99]

In a society of increasing complexity, in which the process of collecting, exchanging, and discarding of information is proliferating at an unparalleled speed, and in which success is measured in terms of one's ability to process larger chunks of information, it is easy to see why biologists might come to see the same forces at work in nature.

The story of creation is being retold. This time around, nature is cast in the image of the computer and the language of cybernetics, the operating tools of the biotechnical age. With both the computer and living organisms, time becomes the primary consideration. Each succeeding generation of computers is more adept at processing increasing amounts of information in shorter periods of time. Coincidentally, the biologists are coming to see a similar pattern of development in nature. According to the new temporal theory, each succeeding species in the evolutionary chain is more adept at processing increasing amounts of information in shorter periods of time. It's not hard to understand, then, why the first generation raised in a fully computerized society will come to accept so readily the new concept of nature that is emerging. They will grow up using the computer to organize their entire environment. Is it any wonder, then, that they will come to believe that nature itself is organized by the same set of assumptions and procedures they themselves are using when they manipulate it?

To sum up, cybernetics is the organizing framework for the coming age, the computer is the organizing mechanism, and living tissue is the organizing material. Cybernetics fuses the computer and living organisms into a single technological context by providing a common language.

It is instructive to recall that even the initial discovery of the double-helix structure of DNA by Sir Francis Crick and James Watson was explained in the language of the computer sciences. M.I.T.'s Joseph Weizenbaum, an authority on the role of the computer in modern society, says that from the very beginning of the revolution in molecular biology the computer provided the appropriate metaphor and computer language provided the appropriate explanation for understanding how biological processes function.

> The results announced by Crick and Watson fell on a soil already prepared by the public's vague understanding of computers, computer circuitry, and information theory. . . . Hence it was easy for the public to see the "cracking" of the genetic code as an unraveling of a computer program, and the discovery of the double-helix structure of the DNA molecule as an explication of a computer's basic wiring diagram.[100]

Now that cybernetics has been firmly accepted as the linguistic framework for both living tissue and mechanical processes, the

technological groundwork has been laid for using the computer
to engineer living tissue.

In the second great economic epoch, the computer sciences
and the bioengineering sciences fuse together into a single technological configuration. Already scientists are busy at work preparing for the union. In a cover article appearing in the May
1982 issue of *Discover* magazine, the popular science journal
owned by Time, Inc., scientists sketch out the new vision of the
organic computer.

> On the outside it looks like another ho-hum electronic device,
> perhaps a hand-held calculator. But inside this garden variety
> box lurks an alien computer. In place of the usual green plastic
> boards holding silicone microchips are ultra-thin films of glass
> crusted over with invisible layers of proteins, linked together in
> complex crystal patterns not unlike those of the arctic retreat in
> the movie *Superman.* Within the delicate protein latticework are
> organic molecules, called biochips, that dance at the touch of an
> electric current, winding or unwinding, passing hydrogen atoms
> from one end to the other. . . .
>
> As they shift positions or shapes, the molecules pass along in-
> formation in the manner of ordinary integrated circuits. But be-
> cause they are so tiny and so close together, they can perform a
> calculation in about a millionth the time of today's best chips.
> One more thing: these molecular diodes, transistors, and wires, as
> well as the protein architecture that holds everything together,
> were manufactured by simple *E. coli* bacteria fashioned to do the
> job by genetic engineering. It can almost be said that the com-
> puter is alive.[101]

The meshing of computer and living tissue will result in a new
type of world economy, one made up almost exclusively of biologically engineered utilities. At this stage it's simply impossible
for the human mind to imagine the contours and boundaries,
the appurtenances and processes of such an alien environment,
for we still relate to a world forged in the fires of the age of pyrotechnology.

The age of biotechnology is likely to unfold in three distinct
stages, with knowledge gained in each providing the basis for the

next. The first stage is already well along. Scientists are learning more each day about how to modify genes, insert genes, and change genes. With genetic engineering, biologists are becoming increasingly adept at changing specific characteristics. However, genetic engineering is merely the first manifestation of the age of biotechnology. As the scientists increase their knowledge of how genes function, they will also become increasingly aware of their limited role within the organism. At the same time their attention will turn increasingly to those forces beyond the gene that exercise a controlling influence over it. An understanding of the cybernetic relationship between gene, cell, organism, and environment will lay the basis for the second stage of the age of biotechnology. At this stage, scientists will be able to expand beyond the engineering of genetic characteristics and begin applying engineering design to the construction of entire organisms. Moving from the engineering of simple temporal characteristics to the engineering of entire temporal programs will have to be followed closely by the development of stage three. In fact, stage two and three are likely to overlap, since both are so dependent on each other for their expression. Stage three is the engineering of entire ecosystems and involves the sophisticated programming of systems of biological information contained within other systems of biological information.

The three stages of bioengineering, then, deal with the engineering of the individual characteristics of the organism, the engineering of the organism itself, and the engineering of the entire ecosystem. Engineering on the first level requires knowledge of the genetic code. Engineering on the second and third levels requires knowledge of biological clocks and fields. The unifying language that fuses characteristics, organisms, and ecosystems with genetic codes, clocks, and fields is cybernetics—the language of the biotechnical age.

PART SIX

THE NEW COSMIC MIRROR

Finding a "Natural" Excuse for the Next World Epoch

A student of cosmological history would entertain little doubt as to the inevitable acceptance of a temporal theory of evolution, even if most scientists have yet to "discover" it. Quite simply, humanity is about to view nature in temporal terms because it is now beginning to engineer the entire temporal life span of living things. Bioengineering is the manipulation of the becoming process of living organisms in advance. For the first time, it is possible to envision the "engineering" of the internal biology of an organism at conception so as to control its entire future development. When scientists engineer changes in the genetic code, they are programming the life cycle of the organism before it unfolds. It is now possible to program a new gene into an organism at conception in order to effect a change in the activity of that organism years later. Controlling the future by designing the temporal programs of living organisms is the central dynamic of the age of biotechnology. We are about to engineer the life spans of all living things in advance, and our cosmology is changing, reflecting this fundamental shift in the way we go about organizing the world around us.

Right down the line, the new temporal theory of evolution reads like a cosmic recommendation for bioengineering. To begin with, bioengineering is an attempt to speed up the conversion of living material into economic utilities in order to secure an ever-expanding growth curve. Molecular biologist James F. Danielli predicts that he and his colleagues will soon speed up nature's way of doing things to the magnitude of one billion times a year.[1] Bioengineering is an attempt to streamline the living process, to make it more efficient, to improve its performance. It is no accident that just as scientists prepare to artificially speed up the temporal programs of living things, they are beginning to advance a theory of life in which evolutionary development follows a trajectory of ever more efficient temporal readjustments.

This is only the beginning of a long string of convenient "coincidences." For example, now that we are engineering organisms to make them compatible with an environment we have created, we contend that organisms are continually re-engineering themselves as they develop in order to comport with their own changing environments. Moreover, because we are now able to fundamentally change an organism's characteristics virtually instantaneously, we have come to believe that fundamental changes in nature occur suddenly and rapidly. A new generation of scientists is advancing the idea of punctuated equilibrium, which argues that basic biological changes in nature occur rapidly and suddenly, not slowly and piecemeal, as Darwinists have long contended. It is also interesting to note that according to the punctuated-equilibrium theory new species develop in total isolation from the parent stock. Of course, in the biotechnical age such will also be the case. Scientists will be able to create new species by manipulating genetic material in a totally isolated laboratory environment.

Finally, according to the new temporal theory, evolution is the steady advance in information processing. Each species up the evolutionary chain is supposedly better able to control greater amounts of information. Coincidentally, increased control over information processing happens to be exactly what bioengineering is all about.

Once again, we have managed to construct a concept of nature that is remarkably sympathetic to the way we happen to be managing nature. Kenneth M. Sayre accurately describes humanity's newest "rationale" for the manipulation of nature when he writes:

> Human beings ... excel in the acquisition of information, and also in versatility of information processing. . . . Since superiority in information gathering and processing amounts to superior adaptive capacities, this accounts for human dominance over other kinds.[2]

Suddenly the old Darwinian notion of "survival of the fittest" is replaced by the idea of "survival of the best informed." Mental acumen, not brute force, becomes the key to evolutionary advancement. Since human beings are the best information processors of all, there is no question that we will extend the evolutionary process to the next stage by engineering new biological systems in the laboratory.

We are on the edge of a new era, one in which life itself will be programmed by engineering design principles. A quarter of a century ago, Norbert Wiener prophesied the coming age of biotechnology. His vision provided an operational approach for the engineering of life as well as a cosmological justification for going ahead with it: "It is my thesis that the physical functioning of the living individual and the operation of some of the newer communication machines are precisely parallel in their analogous attempts to control entropy through feedback."[3]

That being the case, there is no reason whatsoever why bioengineering shouldn't proceed. After all, if, as Wiener and his protégés in the fields of engineering and biology contend, living organisms and machines closely resemble each other, then bioengineering is just an "amplification" of nature's own operational principles. As such, bioengineering is merely a logical extension of, but hardly a radical departure from, the way nature itself operates.

In every particular, the new temporal theory of evolution seems to suggest that nature has always operated much the same way we are operating when we engineer it in the laboratory. Of course, in a sense, the new cosmology contains a small germ of truth. If nature didn't exhibit "some" of the characteristics we ascribe to it, we would find it impossible to manipulate it the way we are in the laboratory. The problem, once again, is that in our cosmologies we inflate the tiny aspect of nature's reality that we are manipulating at a moment in time into a universal cosmology and then claim that "all" of nature operates in a manner that is congenial with the way we are operating. We continually remake nature to suit our own needs and then conclude that the technological procedures we are using at the time must be similar to the procedures used in constructing the original creation. As we have seen, this is exactly the error that Darwin made in the nineteenth century, when he observed animal breeders producing specific varieties of livestock by artificial selection techniques. He immediately inflated his observation by contending that all of nature developed and continues to operate in a similar fashion.

People's organizing modes have always stretched out until they have filled the entire universe. That's because we need continually to convince ourselves that the way we are going about organizing the world is, in fact, compatible with the way the world itself is organized. Our cosmology provides the blanket of security we need to continue our manipulation without having to entertain doubts as to the efficacy of our actions. By formulating a view of nature that is congenial with the way we are exploiting nature, we are able to lend an air of "cosmic legitimacy" to our technological pursuits. Cosmologies, then, serve as the ultimate form of justification for our day-to-day activity in the world. They allow us to continue the fiction that our behavior conforms with the "natural order of things."

The New Immortality

As we have seen, every human epoch can be characterized by the way it organizes the world around it. Each organizing style, in turn, represents humanity's best attempt to overcome the limits imposed by time and space. The goal is always the same. We organize to perpetuate ourselves, and our dream is to organize ourselves so well that we will be able to overcome our earthly sojourn and experience some measure of immortality. The dream of immortality pushes the human race onward to new organizational feats, each more nimble, more quick, but none nimble enough or quick enough to allow us to jump over time and space into the world beyond the senses. Still, humanity is never without a new scheme, always in the belief that sooner or later the human spirit will finally triumph over its own bodily fate. So every epoch has had its own organizational vehicle, although the road being traveled along is always the same. The path to immortality is littered with the organizational plans of the ages. Every one of these organizational relics expressed people's hope for an ultimate victory over the forces of finality. In the Middle Ages, skilled craftsmen and chemists dreamed of turning lead into gold and of mixing chemicals with fire to discover the secret elixir that would guarantee everlasting life. With the dawn of the Industrial Age, people's dreams turned from alchemy to perpetual motion. Generations of inventors and machinists gave over their lives and their fortunes in their quest to build the perfect machine; one that would run by itself, be totally self-contained, and thus live forever.

Today we are equipping ourselves with a new organizing vehicle, hoping to complete the journey our ancestors never finished. The age of biotechnology incorporates its own unique vision of immortality. The ultimate organizational goal of the biotechnological revolution is best expressed in a popular television series seen by millions of people throughout the world. In *Star Trek*, the starship *Enterprise* has a special room called the transporter room. Personnel wishing to leave the starship do so by entering the transporter room. The transporter itself is a sophisticated computer that acts as a "matter/energy scrambler."[4] According to Captain Kirk, the transporter "converts" matter temporarily into energy, beaming that energy to a fixed point. In other words, the human body is transformed into billions of bits of information, which are then sent over space just like elec-

tronic impulses. The information is then reassembled at its desti-
nation, restoring the body to its original form. The ship's doctor,
McCoy, points out that people can even be "suspended in transit
until a decision is reached to rematerialize them."[5] The transporter ". . . retains the memory of the original molecular structure of everyone and everything passing through it,"[6] allowing it
to turn information from matter to energy and back to matter
again.

The ability to reduce all biological systems to information and
then to use that information to overcome time and space limitations is the ultimate dream of the biotechnologists. The vision of
the transporter room provides a grand organizational goal to aspire to, one as seductive to a molecular biologist as perpetual
motion was to an industrial engineer and alchemy to a medieval
metallurgist. Of course, many biotechnologists might well take
exception to this notion of ultimate goals. Perhaps it is only because they have not yet thought through the full implication of
where their work is leading them. One scientist has, however.
Not surprisingly, it is Norbert Wiener, the man responsible for
transporting cybernetic principles from engineering to biology.
It is appropriate to quote Mr. Wiener on the subject of goals, for
his organizing vision has been largely responsible for laying the
foundation for the age of biotechnology.

Wiener starts off with the assumption that all living things are
really "patterns that perpetuate themselves," and that "a pattern is a message and may be transmitted as a message."[7] Wiener
then asks:

> What would happen if we were to transmit the whole pattern of
> the human body . . . so that a hypothetical receiving instrument
> could re-embody these messages in appropriate matter, capable of
> continuing the processes already in the body and the mind. . . ?[8]

Wiener concludes that "the fact that we cannot telegraph the
pattern of a man from one place to another seems to be due to
technical difficulties . . ." but he hastens to add that "the idea itself is highly plausible."[9] It is likely that the biotechnologists will
never completely work out those "technical difficulties," but

that's of little matter. The mechanics of the Industrial Age were never able to work out the "technical difficulties" of producing perpetual motion, but there is no doubt that the vision provided a distant goal for their journey.

What is important is that the new organizing mode and its cosmological companion offer the hope of immortality somewhere in the future. The idea of overcoming time and space, and of preserving life over eons of time, is a familiar beacon. It illuminates the same far-off port that every other organizing mode and cosmology has set sail for: the port of everlasting immortality.

The new cosmologists believe that information is the key to immortality. Information, they contend, is impervious to the ravages of time. It can be stored and passed along. It accumulates. It is of this world but does not die with the flesh. Just as a Christian might contend that the body is merely a temporary encasement for the everlasting spirit that resides in it, the new cosmologists would contend that the body is merely a temporary encasement for the information that gives rise to it. Carl Sagan once suggested that transmitting the entire information code of a cat to some superior intelligence somewhere else in the cosmic theater would be the same as sending the cat itself. Physical organisms live and die. The information that makes up these temporary forms does not. Even today, biologists are storing the germ plasm of rare plants and animals in special gene banks. They hope to reconstruct the living forms sometime in the future by learning how to set in motion the information contained in the genetic codes.

Succeeding generations will seek after the information of living processes the way past generations sought the elixir and perpetual motion. Controlling larger and more complex stores of biological information will be the immortality symbol of the coming age. For our heirs, the path to eternal life will be paved with ream upon ream of informational readouts.

The New Self-containment

Developments in the information sciences and in biology have greatly expanded humanity's temporal horizon. As in past periods of history, greater anticipation of the future has heightened the tension between people's desire for self-sufficiency on the one hand and our increasing awareness of our relationship to all other living things on the other. In the Darwinian cosmology, all organisms are related by their common biological ancestry, but human beings also claim a separate and unique status by dint of their superior position at the top of the biological hierarchy. Even though humanity enjoys a privileged status, it resents being dependent on and obligated to all the "lower forms" of life for its own existence. Darwin's cosmology resolves the tension between people's desire for invulnerability and nature's demand for dependence by claiming that even though each organism fights for its own self-sufficiency, it inadvertently advances the common good by contributing its improved genes to the evolutionary process. Thus humanity can expiate its sense of guilt by believing that its ruthless drive for self-mastery actually serves to enhance nature's grand operating design. Like all past cosmologies, then, Darwin's provided humanity with a rationale by which it could continue to exploit the rest of the living kingdom with impunity.

In the new temporal cosmology, cybernetics replaces natural selection, providing a new resolution of the conflict between the desire for self-containment and the increased awareness of dependency. Under the new schema, all living things are related to one another in the evolutionary hierarchy in that each served as a source of information for the species that superseded it. Each species is "better informed" because it has been able to incorporate the information of past species into its own temporal program. Every species, therefore, is obligated to all earlier forms of life for adding to the store of information at its disposal. At the same time, every organism during its lifetime depends on the exchange of information with other living things to ensure its self-perpetuation.

While cybernetics establishes a new level of interrelatedness between all living things, it also provides humankind with a way to escape the bonds of debt and dependency that it evokes. Cybernetics replaces Darwin's notion of maximizing self-interest with the idea of maximizing self-organization. According to the new temporal cosmology, self-sufficiency is no longer achieved by way of natural selection, but rather by way of negative and positive feedback. Every organism seeks to maximize its own self-organization by exchanging information with its environment. While each species seeks only its self-organization, in the process it generates new bits of information, which are the source of further evolutionary development. Every new evolutionary advance, in turn, increases the overall complexity of the system, further integrating all the information into a richer labyrinth of relationships.

Cybernetics provides humanity with a new rationale for the continued manipulation of the environment. In an age steeped in the information mystique, people can take comfort in the belief that their own efforts to generate and control greater stores of information not only advance their self-organization but also contribute to the strengthening of all relationships in nature by increasing the level of interaction, interconnection, and synchronization in the system as a whole.

The New Desacralization

Every cosmology has a way of making the existing mode of organization appear completely benign. It is through the process of desacralization that the defanging takes place. Every time the human family changes the way it goes about organizing the world, it finds it necessary to sever any sense of empathetic association it might feel toward the objects of its assimilation. It's much more difficult to exploit something you identify with, so desacralization serves as a kind of psychic ritual by which human beings deaden their prey, preparing it for consumption.

Darwin's world was populated by machine-like automata.

Nature was conceived as an aggregate of standardized, interchangeable parts assembled into various functional combinations. If one were to ascribe any overall purpose to the entire
operation, it would probably be that of increased production
and greater efficiency with no particular end in mind.

The new temporal theory of evolution replaces the idea of life
as mere machinery with the idea of life as mere information. By
resolving structure into function and reducing function to information flows, the new cosmology all but eliminates any remaining sense of species identification. Living things are no longer
perceived as carrots and peas, foxes and hens, but as bundles of
information. All living things are drained of their aliveness and
turned into abstract messages. Life becomes a code to be deci-
phered. There is no longer any question of sacredness or inviola-
bility. How could there be when there are no longer any
recognizable boundaries to respect? Under the new temporal
theory, structure is abandoned. Nothing exists at the moment.
Everything is pure activity, pure process. How can any living
thing be deemed sacred when it is just a pattern of information?

By eliminating structural boundaries and reducing all living
things to information exchanges and flows, the new cosmology
provides the proper degree of desacralization for the bioengineering of life. After all, in order to justify the engineering of living material across biological boundaries, it is first necessary to
desacralize the whole idea of an organism as an identifiable, discrete structure with a permanent set of attributes. In the age of
biotechnology, separate species with separate names gradually
give way to systems of information that can be reprogrammed
into an infinite number of biological combinations. It is much
easier for the human mind to accept the idea of engineering a
system of information than it is for it to accept the idea of engineering a dog. It is easier still, once one has fully internalized the
notion that there is really no such thing as a dog in the traditional sense. In the coming age it will be much more accurate to
describe a dog as a very specific pattern of information unfolding over a specific period of time.

Life as information flow represents the final desacralization of
nature. Conveniently, humanity has eliminated the idea of fixed

biological borders and reduced matter to energy and energy to information in its cosmological thinking right at the very time that bioengineers are preparing to cut across species boundaries in the living world.

The New Ethics

Civilization is experiencing the euphoric first moments of the next age of history. The media are already treating us to glimpses of a future where the engineering of life by design will be standard operating procedure. Even as the corporate laboratories begin to dribble out the first products of bioengineering, a subtle shift in the ethical impulse of society is becoming perceptible to the naked eye. As we begin to reprogram life, our moral code is being similarly reprogrammed to reflect this profound change in the way humanity goes about organizing the world. A new ethics is being engineered, and its operating assumptions comport nicely with the activity taking place in the biology laboratories.

Eugenics is the inseparable ethical wing of the age of biotechnology. First coined by Charles Darwin's cousin Sir Francis Galton, eugenics is generally categorized in two ways, negative and positive. Negative eugenics involves the systematic elimination of so-called biologically undesirable characteristics. Positive eugenics is concerned with the use of genetic manipulation to "improve" the characteristics of an organism or species.

Eugenics is not a new phenomenon. At the turn of the century the United States sported a massive eugenics movement. Politicians, celebrities, academicians, and prominent business leaders joined together in support of a eugenics program for the country. The frenzy over eugenics reached a fever pitch, with many states passing sterilization statutes and the U.S. Congress passing a new immigration law in the 1920s based on eugenics considerations. As a consequence of the new legislation, thousands of American citizens were sterilized so they could not pass on their "inferior" traits, and the federal government locked its doors to certain immigrant groups deemed biologically unfit by then-existing eugenics standards.

While the Americans flirted with eugenics for the first thirty years of the twentieth century, their escapades were of minor historical account when compared with the eugenics program orchestrated by the Nazis in the 1930s and '40s. Millions of Jews and other religious and ethnic groups were gassed in the German crematoriums to advance the Third Reich's dream of eliminating all but the "Aryan" race from the globe. The Nazis also embarked on a "positive" eugenics program in which thousands of S.S. officers and German women were carefully selected for their "superior" genes and mated under the auspices of the state. Impregnated women were cared for in state facilities, and their offspring were donated to the Third Reich as the vanguard of the new super race that would rule the world for the next millennium.

Eugenics lay dormant for nearly a quarter of a century after World War II. Then the spectacular breakthroughs in molecular biology in the 1960s raised the specter of a eugenics revival once again. By the mid-1970s, many scientists were beginning to worry out loud that the potential for genetic engineering might lead to a return to the kind of eugenics hysteria that had swept over America and Europe earlier in the century. Speaking at a National Academy of Science forum on recombinant DNA, Ethan Signer, a biologist at M.I.T., warned his colleagues that

> this research is going to bring us one more step closer to genetic engineering of people. That's where they figure out how to have us produce children with ideal characteristics. . . . The last time around, the ideal children had blond hair, blue eyes and Aryan genes.[10]

The concern over a re-emergence of eugenics is well founded but misplaced. While professional ethicists watch out the front door for telltale signs of a resurrection of the Nazi nightmare, eugenics doctrine has quietly slipped in the back door and is already stealthily at work reorganizing the ethical priorities of the human household. Virtually overnight, eugenics doctrine has gained an impressive if not an impregnable foothold in the popular culture.

Its successful implantation into the psychic life of civilization is attributable to its going largely unrecognized in its new guise. The new eugenics is commercial, not social. In place of the shrill eugenic cries for racial purity, the new commercial eugenics talks in pragmatic terms of increased economic efficiency, better performance standards, and improvement in the quality of life. The old eugenics was steeped in political ideology and motivated by fear and hate. The new eugenics is grounded in economic considerations and stimulated by utilitarianism and financial gain.

Like the ethics of the Darwinian era, the new commercial eugenics associates the idea of "doing good" with the idea of "increasing efficiency." The difference is that increasing efficiency in the age of biotechnology is achieved by way of engineering living organisms. Therefore, "good" is defined as the engineering of life to improve its performance. In contrast, not to improve the performance of a living organism whenever technically possible is considered tantamount to committing a sin.

For example, consider the hypothetical case of a prospective mother faced with the choice of programming the genetic characteristics of her child at conception. Let's assume the mother chooses not to have the fertilized egg programmed. The fetus develops naturally, the baby is born, the child grows up, and in her early teenage years discovers that she has a rare genetic disease that will lead to a premature and painful death. The mother could have avoided the calamity by having that defective genetic trait eliminated from the fertilized egg, but she chose not to. In the age of biotechnology, her choice not to intervene might well constitute a crime for which she might be punished. At the least, her refusal to allow the fetus to be programmed would be considered a morally reprehensible and irresponsible decision unbefitting a mother, whose duty it is always to provide as best she can for her child's future well-being.

Proponents of human genetic engineering contend that it would be irresponsible not to use this powerful new technology to eliminate serious "genetic disorders." The problem with this

argument, says *The New York Times* in an editorial entitled
"Whether to Make Perfect Humans," is that "there is no discernible line to be drawn between making inheritable repairs of
genetic defects, and improving the species."[11] The *Times* rightly
points out that once scientists are able to repair genetic defects,
"it will become much harder to argue against adding genes that
confer desired qualities, like better health, looks or brains."[12]

Once we decide to begin the process of human genetic engineering, there is really no logical place to stop. If diabetes, sickle
cell anemia, and cancer are to be cured by altering the genetic
makeup of an individual, why not proceed to other "disorders":
myopia, color blindness, left-handedness? Indeed, what is to
preclude a society from deciding that a certain skin color is a disorder?

As knowledge about genes increases, the bioengineers will inevitably gain new insights into the functioning of more complex
characteristics, such as those associated with behavior and
thoughts. Many scientists are already contending that schizo-
phrenia and other "abnormal" psychological states result from
genetic disorders or defects. Others now argue that "antisocial"
behavior, such as criminality and social protest, are also examples of malfunctioning genetic information. One prominent
neurophysiologist has gone so far as to say, "There can be no
twisted thought without a twisted molecule."[13] Many sociobiol-
ogists contend that virtually all human activity is in some way
determined by our genetic makeup, and that if we wish to
change this situation, we must change our genes.

Whenever we begin to discuss the idea of genetic defects, there
is no way to limit the discussion to one or two or even a dozen
so-called disorders, because of a hidden assumption that lies behind the very notion of "defective." Ethicist Daniel Callahan
penetrates to the core of the problem when he observes that "behind the human horror at genetic defectiveness lurks ... an
image of the perfect human being. The very language of 'defect,'
'abnormality,' 'disease,' and 'risk,' presupposes such an image, a
kind of proto-type of perfection."[14]

The idea of engineering the human species is very similar to
the idea of engineering a piece of machinery. An engineer is

constantly in search of new ways to improve the performance of a machine. As soon as one set of imperfections is eliminated, the engineer immediately turns his attention to the next set of imperfections, always with the idea in mind of creating a perfect piece of machinery. Engineering is a process of continual improvement in the performance of a piece of machinery, and the idea of setting arbitrary limits to how much "improvement" is acceptable is alien to the entire engineering conception.

The question, then, is whether or not humanity should "begin" the process of engineering future generations of human beings by technological design in the laboratory. What is the price we pay for embarking on a course whose final goal is the "perfection" of the human species? How important is it that we eliminate all the imperfections, all the defects? What price are we willing to pay to extend our lives, to ensure our own health, to do away with all the inconveniences, the irritations, the nuisances, the infirmities, the suffering, that are so much a part of the human experience? Are we so enamored with the idea of physical perpetuation at all costs that we are even willing to subject the human species to rigid architectural design?

With human genetic engineering, we get something and we give up something. In return for securing our own physical well-being we are forced to accept the idea of reducing the human species to a technologically designed product. Genetic engineering poses the most fundamental of questions. Is guaranteeing our health worth trading away our humanity?

People are forever devising new ways of organizing the environment in order to secure their future. Ethics, in turn, serves to legitimize the drive for self-perpetuation. Any organizing activity that a society deems to be helpful in securing its future is automatically blessed, and any activity that undermines the mode of organization a society uses to secure its future is automatically damned. The age of bioengineering brooks no exception. In the years to come a multitude of new bioengineering products will be forthcoming. Every one of the breakthroughs in bioengineering will be of benefit to someone, under some circumstance, somewhere in society. Each will in some way appear to advance the future security of an individual, a group, or society as a whole. Eliminating a defective gene trait so that a child won't die prematurely; engineering a new cereal crop that can feed an expanding population; developing a new biological source of energy that can fill the vacuum as the oil spigot runs dry. Every one of these advances provides a modicum of security against the vagaries of the future. To forbid their development and reject their application will be considered ethically irresponsible and inexcusable.

Bioengineering is coming to us not as a threat but as a promise; not as a punishment but as a gift. We have already come to the conclusion that bioengineering is a boon for humanity. The thought of engineering living organisms no longer conjures up sinister images. What we see before our eyes are not monstrosities but useful products. We no longer feel dread but only elated expectation at the great possibilities that lie in store for each of us.

How could engineering life be considered bad when it produces such great benefits? Engineering living tissue is no longer a question of great ethical import. The human psyche has been won over to eugenics with little need for discussion or debate. We have already been convinced of the good that can come from engineering life by learning of the helpful products it is likely to spawn.

As in the past, humanity's incessant need to control the future in order to secure its own well-being is already dictating the ethics of the age of biotechnology. Engineering life to improve humanity's own prospects for survival will be ennobled as the highest expression of ethical behavior. Any resistance to the new technology will be castigated as inhuman, irresponsible, morally reprehensible, and criminally culpable.

The New Politics

The age of biotechnology will effect a fundamental change in how we govern ourselves. Gradually, the orthodox notion of rule as a spatial conception will give way to the idea of governance as a temporal conception. We are already at the beginning stages of a historic transformation from "rule by territory" to "management by system." The electronics revolution and subsequent advances in the information sciences and in communications allow people to penetrate geographical borders with the same ease that we now penetrate biological borders. Information webs are now capturing human populations, bringing them under the control of a new type of imperialism, one that can sweep across space and invade the interior landscape of any region of the globe with impunity. With satellite communications people find themselves beholden to systems that extend beyond the geographical boundaries of the state. Transnational corporations and other economic networks are now spanning the globe, wresting populations from traditional geographic loyalties. The result is that the nation-state is gradually being subsumed by new "patterns" of social discourse. Space is no longer a limiting factor. It no longer separates and divides people as effectively as in the past. While the nation-state is still the dominant reality and is likely to remain so for quite a while, its eventual demise in the far-distant future is already dimly perceived.

The new imperialism is temporal and cybernetic. The key to political power in the coming age is the effective exercise of control over the information systems for the many processes that connect living organisms with one another and their environments. In cybernetics language, positive feedback replaces imperialism and negative feedback replaces colonialization. The first is the means by which a system expands its temporal control, bringing more and more information processes under its dominion. The second is the means by which a system manages the new temporalities it has imperialized. With the dawn of the new age, politics begins to shift gradually from a spatial to a temporal plane and from the idea of controlling geographical locations to the idea of controlling patterns and processes. The most important patterns and processes are those that make up human life itself. It is here that the politics of the biotechnical age will be most intimately expressed.

Thoughout history some people have always controlled the

futures of other people. The politics of the biotechnical age are no exception. Every new biotechnical discovery will affect the power that some people exercise over others. The exercise of political power in the coming age will raise terrifying questions. For example, whom would we trust with the decision of what is a good gene that should be added to the gene pool and what is a bad gene that should be eliminated? Would we trust the federal government? the corporations? the university scientists? a group of our peers? From this perspective, few of us are able to point to any institution or group of individuals we would entrust with decisions of such import. However, if instead of the above question, we were asked whether we would sanction new bioengineering products that could provide new sources of food and energy and enhance the physical and mental well-being of people, we would not hesitate for a moment to add our support. Yet all biologically engineered products require that someone make a decision about which genes should be engineered and which genes should be done away with. How, then, is it possible for people to be leery of trusting anyone with authority over genetic technology and at the same time be in favor of the development of the technology itself? The answer is to be found in the nature of the question. The first question deals with power in a vacuum. No one is willing to blindly hand over power over his own future—especially when it involves the engineering of life itself. The second question, however, deals with the exchange of power for material security. Virtually everyone is willing to give up some of his own power in return for being guaranteed some measure of material security in return. Exchanging power for security is the nature of politics.

The population at large has already made its intentions clear so far as the politics of the biotechnical age are concerned. An unwritten contract of sorts has already been agreed upon, and without the wrangling and negotiations that usually accompany the struggle over exchange of power. The power to control the future biological design of living tissue has been signed over to the scientists, the corporations, and the state without ceremony. In return, all that is being asked for are useful products that will enhance human survival and provide for the general well-being.

At first blush, the bargain appears a good one. Biotechnology has much to offer. But, as with other organizing modes throughout history, the final costs have not yet been calculated. Granting power to a specific institution or group of individuals to determine a better-engineered crop or animal or a new human hormone seems such a trifle in comparison with the potential returns. It is only when one considers the lifetime of the agreement that the full import of the politics of the biotechnological age becomes apparent.

Power is the exercise of control over the future. Throughout history humanity has been gradually increasing its power by extending its control over broader temporal horizons. Today, the ultimate exercise of power is within grasp: the ability to control the future of all living things by engineering their entire life process in advance, making them a hostage of their own architecturally designed blueprints. Bioengineering represents the power of authorship. Never before in history has such complete power over life been a possibility. The idea of imprisoning the entire life cycle of an organism by simply engineering its organizational blueprint at conception is truly awesome.

In these early stages of the age of biotechnology such power, though formidable, appears so far removed from any potential threat to the human physiology as to be of little concern. We are more than willing to allow the rest of the living kingdom to fall under the shadow of the engineering scalpel, as long as it produces some concrete utilitarian benefit for us. We are even willing to subject parts of our anatomy to bioengineering if it will enhance our physical and mental health. The problem is that biotechnology has a beginning but no end. Cell by cell, tissue by tissue, organ by organ, we give up our bodies as we give up our political power, a piece at a time. In the process, each loss is compensated for with a perceived gain until there is nothing more to exchange. It is at that very point that the cost of our agreement becomes visible. But it is also exactly at that point that we no longer possess the very thing we were so anxious to preserve: our own security. In the decades to come, we humans will barter ourselves away, a piece at a time, in exchange for some measure of temporary well-being. In the end, the security we fought so hard to preserve will have disappeared forever. Thanks to bioengineering, we will finally have been extricated from the great burden of human history, the unremittent need to anticipate and secure our own future. Security will no longer be our concern, because we will no longer control any measure of our own destiny. Our future will be determined at conception. It will be programmed into our biological blueprint.

The Chameleon Cosmology

Once again: Every cosmology reinforces the organizing mode it serves. So too with the temporal theory of evolution. Like its predecessors, it provides an overarching rationale for humanity's newest organizing exploit. According to the new theory, evolution is the creative advance in the store of knowledge. Each species is "better informed" than its predecessor and thus better equipped to anticipate and control its future. This being the case, human beings are merely conforming to the processes of nature every time they add to their own store of knowledge. Therefore, to place limits on the pursuit of knowledge would appear to be "unnatural" and therefore undesirable. If evolution is the increase in knowledge, then humanity is performing its proper role in the cosmic scheme by its relentless drive to know more and more in order to anticipate and control its own future.

In the new cosmology it is information that is elevated and made the primary object of our attention. Knowledge for knowledge's sake becomes the *raison d'être* of the coming age. Gone are the restraints imposed on the pursuit of knowledge by earlier cosmologies. There is no more warning of Pandora's box or of biting into the apple from the Tree of Knowledge, because we no longer fear "a source" of knowledge beyond our reach. No longer beholden to the gods, people are convinced that knowledge is not something that is handed down, but rather, is something that evolves.

According to the new temporal theory, the evolution of information is tantamount to the evolution of life. With this newly constructed cosmology, Prometheus can now end his long agony and Adam and Eve can return to the Garden. The true scientist need no longer fear his quest for knowledge or make paltry appeals to the ghost of Galileo for his every scientific foray. Since evolution is the accumulation of more and more information, the scientist, by his every thought and action, is advancing the development of life on earth.

With its new, more exalted status, knowledge changes its complexion. It is no longer viewed as "a discovery of facts," but rather as "an ongoing creative process." In all earlier periods of history, knowledge came slowly. For that reason, each new insight was enshrined and closely guarded. It was elevated and made timeless. It became something to live by. It lasted because great lapses of time occurred between insights. That's no longer the case. What we know today is quickly eclipsed by what we will know tomorrow. Thus, we can no longer tolerate timeless truths and ironclad laws. By their very nature, timeless truths impose boundaries, and it is for that reason that they are now dispensable. Timeless truths and ironclad laws tell us what is not possible. They establish upward limits to what we can do. In every cosmology they serve as a notice as to how far it is possible to go.

Today, everything changes so fast that it is necessary to construct a cosmology in which change itself is honored as the only timeless truth. By reinterpreting nature as the evolution of information, humanity achieves this end. Nature is no longer seen as a set of restraints, but rather as a process of creative advance. In this new scheme of things even the laws of science lose their potency. They are no longer seen as ironclad truths but merely as convenient instruments to advance the information process. Nothing is considered permanent except the ongoing process of information gathering and processing.

It is important, at this point, to note the subtle but profound shift in conceptual thought that is occasioned by using the terms "knowledge" and "information" interchangeably. Knowledge is losing its special status and is increasingly being subsumed under the category of information. The distinction between knowledge and information is becoming blurred. Today, "to be knowledgeable" and "to be informed" have come to mean virtually the same thing. This is a revolution in the history of conceptual thought. Knowledge has been reduced to information, and as a result, it has lost its sacred status. Knowledge is no longer seen as something absolute and permanent that we discover, something that exists unaffected and untouched by the passage of time. By changing the meaning of knowledge, so that "to know" becomes equivalent to "being informed," we saturate knowledge with temporality. "To be informed" means to be aware of changing conditions. Being informed requires a constant updating. It is an ongoing process of anticipation and accommodation to changes going on in the environment. As knowledge becomes more and more synonymous with information, it loses its sense of permanence, its timeless status. To be knowledgeable today is to be continually aware of the changes going on around us and to be able to adjust accordingly. This is a far different sense of knowledge from the one held by previous generations, who, when they used the term, thought of the discovery of permanent facts that existed in a kind of timeless vacuum. Knowledge gathering is now information gathering, and information gathering is the processing of changing conditions over time. With this linguistic metamorphosis, the door is opened to a thorough reinterpretation of existence as pure process devoid of any kind of ultimate, unchanging frame of reference.

If evolution is the processing of greater stores of information, as many now contend, then bioengineering will be seen as the inevitable next step in evolutionary history. Bioengineering reorganizes existing information programs and creates new ones. It will shuffle information back and forth between organisms and will eventually create new systems of information as well. Certainly it can be said that no other human undertaking in history has ever been so able to add significantly to the store of new biological information as the age of biotechnology will. Conveniently, humanity has now come to view nature as the creation of more and more information at the very time that it has embarked on a course to create its own engineered nature with the new "information" at its disposal.

The temporal theory offers much more than a convenient rationale. It delineates humanity's new responsibility. One hundred years after Thomas Huxley's eloquent defense of Darwin's theory, his grandson Julian Huxley takes up the banner of the emerging cosmology. Humanity, itself the product of evolutionary creativity, is now obligated, says Huxley, to continue the creative process by becoming the architect for the future development of life. *Homo sapiens'* destiny, he contends, is to be "the sole agent of further evolutionary advance on the planet."[15]

Humanity's new responsibility is indeed awesome. All Darwin asked of people was that they compete for their own life. The new cosmology asks people to be the creator of life itself. Huxley surveys the future and comes to the conclusion that humanity has no other choice but to become the "business manager for the cosmic process of evolution." In this new way of thinking humanity does not choose bioengineering. It is thrust upon it by the forces of nature. It is the next stage in the evolutionary process. It is the inevitable result of "the creative advance in knowledge" that began with the emergence of the first organism. From the very beginnings of life on, each organism has striven to enlarge its informational domain, to become "better informed." That the human mind has now become so well informed that it could actually conceive of using the vast amount of information at its disposal to engineer life is itself a testimonial to the entire evolutionary process.

Therefore, if one accepts the new temporal theory, one has little choice but to accept bioengineering as well. Not to do so would appear to violate the very process of evolutionary development. By the new cosmological thinking, bioengineering is not something artificially superimposed on nature but something spawned by nature's own ongoing evolutionary process. It is, in effect, the next stage in the evolutionary process. Any effort, therefore, to resist bioengineering would in the end be futile and self-defeating because it would fly in the face of what is "natural."

The Final Twist of the Screw

Plato, St. Thomas Aquinas, Charles Darwin ... these were not
evil men. Their cosmologies were not the product of intrigue.
These learned gentlemen were merely trying to express, as best
they could, the workings of nature. They truly believed that
their formulations were an act of discovery, an unmasking of the
universal scheme of things. They sought the truth and, to a man,
believed that it existed somewhere outside themselves. They
were convinced that their cosmologies were an accurate description of the way the world was.

Now, for the first time, humanity is becoming aware of its own
history of self-deception. A new generation of scholars is exposing the past to the full scrutiny of social reconstruction. The
great figures in history are being looked at in a new light. Questions are being asked about their motivations, cultural biases,
and environmental conditioning. The grand cosmological ideas
are being reassessed as well. Scholars are attempting to uncover
the complex technological, economic, social, and political factors
that influence and provide ballast for the great ideas of history.

It is only when consideration turns to the present and future
that there exists an unwillingness to apply the same criteria of
assessment. We are more than willing, even anxious, to expose
the self-deception of the past, but not our own. That is because
humanity always needs to believe that its behavior conforms to
the state of nature. Our generation is no different. We can't afford to believe that our actions have no basis in the larger
scheme of things. Therefore, we continue to conjure up cosmologies that comport nicely with our own behavior in the world.
Some futurists think this is all about to change. They point to
the emergence of a whole new vocabulary of words and terms as
proof of sorts that the self-deception that has guided our cosmologies over the millennia is about to be expurgated once and for
all. For example, they point out that the idea of an "objective"
reality is giving way to the idea of a "perspective" reality. The
idea that future states are subject to ironclad laws of causality is
giving way to the idea that the future is a trajectory of "creative

possibilities." The idea of "deterministic outcomes" is being replaced with the idea of "likely scenarios." The idea of "permanent truths" is being replaced with the idea of "useful models." Many philosophers and scientists are convinced that this abrupt change in vocabulary signals a departure from the long-existing hubris by which humanity has cast itself as "the measure of all things." Quite the contrary; the new language is not an expression of humility. The belief that there are no ironclad truths or some objective reality that human beings can discover does not mark the end of the great self-deception that has long plagued humanity, but only the beginning of a new chapter.

Where Darwin introduced history into his cosmology, our generation is introducing psychology, and by so doing, it is changing the form and context of the deception. Ensuing generations will be more than willing to cast aside the idea of an independent historical reality existing outside themselves to which their behavior need conform. They will be convinced that the world exists inside themselves, that they conceptualize it with their minds, fashion it with their knowledge, and orchestrate it with their energy. Our children will thank us for releasing them from the shackles of a self-imposed historicity. They will march into the future, not out of a sense of historical determinism, but rather from a sense of "expanding consciousness." They will gladly acknowledge that their view of the world is a reflection of their state of mind. In fact, they will argue that it could not be otherwise. After all, if all nature is viewed as an ever-expanding state of consciousness, then humanity's own consciousness doesn't describe nature, it embodies it.

In this context, the new cosmological vocabulary takes on a Janus face. At first glance, terms like "perspective," "scenarios," "models," "creative possibilities" appear to signal a newfound awareness by humanity of its own limitations, of its inability ever to fully grasp or comprehend the truths of the universe. Not so. It is not humility that animates the new cosmological jargon but bravado. When we take a closer look, the new vocabulary suddenly takes on an entirely new appearance, at once menacing and intoxicating. Perspectives, scenarios, models, creative possibilities. These are the words of authorship, the words of a creator, an architect, a designer. Humanity is abandoning the idea that the universe operates by ironclad truths because it no longer feels the need to be constrained by such fetters. Nature is being made anew, this time by human beings. We no longer feel ourselves to be guests in someone else's home and therefore obliged to make our behavior conform with a set of pre-existing cosmic rules. It is our creation now. We make the rules. We establish the parameters of reality. We create the world, and because we do, we no longer feel beholden to outside forces. We no longer have to justify our behavior, for we are now the architects of the universe. We are responsible to nothing outside ourselves, for we are the kingdom, the power, and the glory for ever and ever.

Today, our biotechnical arts merely imitate nature. Tomorrow, they will subsume it. Our children will be convinced that their creations are of a far superior nature to those from which they were copied. They will be the algenists. They will view all of nature as a computable domain. They will redefine living things as temporal programs that can be edited, revised, and reprogrammed into an infinite number of novel combinations. The algenists will change the essence of living things. They will upgrade the performance of living systems. They will program entirely new biological processes. They will seek to transform the living world into a golden treasure trove, a perfectly engineered, optimally efficient state.

Our children will view their imitation of nature as nature. Their art will become their reality, and with that transition, algeny will lose its metaphorical significance, just as alchemy did as it slowly metamorphosed into the scientific world view. Today, algeny is a figure of speech; tomorrow, a way of life.

PART SEVEN

CHOICES

If it were pure illusion, it would be easy to dispense with the new temporal cosmology. But it is not an illusion, it is a deception, and for that reason is much more difficult to contend with. Like all of humanity's past cosmological formulations, this newest construct is extrapolated from reality. The new temporal theory reflects people's emerging organizational relationship with the environment. That organizational relationship is very real indeed. After all, bioengineering works and produces very real consequences in the physical world. The deception lies in the transformation of what is a very real but limited relationship between people and their environment into a universal explanation of the nature of nature. The new temporal theory is not an accurate explanation of nature any more than Darwin's theory of natural selection was. It is merely a fragment of reality that has been amplified and turned into an all-embracing explanation for existence.

Human beings don't like gaps, especially the gap between what we are familiar with and what lies beyond our realm of experience. Cosmologies are our way of convincing ourselves that there is no void, no unimaginable, unthinkable abyss beyond our reach to which we will never be privy and which can never be brought under our domestication. Our cosmologies relieve our great sense of apprehension about what truly lies out there. They assure us that the world does indeed make sense and that it can be understood in the same terms that we understand it.

Our cosmologies protect us. They ease our fears. They make the world seem manageable by projecting it to be a reflection of the world we are managing. If our cosmologies teach us anything, it is that we can feel comfortable only in a world we can fully explain. The problem is that no such world exists. There will always be a yawning gap between the world we experience and the world that exists beyond our experience. Regardless of how hard we try, we will never be able to fill that void. Our cosmologies serve as screens. We use them to hide from the great abyss beyond. We surround ourselves with our cosmologies, and they in turn surround us with a façade of invincibility. We are sure that the world is totally understandable within the organizational framework we create and for that reason remain convinced that we can go on as we are without any fear of harm coming to us.

Our cosmologies sanction our behavior. They tell us that the way we are organizing ourselves and the world around us is in accord with the natural order of things. We bury ourselves in our cosmologies in order to escape from having to peer into the great void that surrounds the tiny edifice we have colonized.

We behave as if the world were made for us, never stopping for a moment to consider the possibility that we might have been made for the world. If the world is ours for the making, then of course we are more than justified in believing that the reality we are organizing is in accord with whatever lies beyond it. But what if such were not the case? What if the opposite were true and we suddenly discovered that we were made for the world? How different our course would be. Nature's relationship to us would no longer be as important as our relationship to nature. Instead of forcing the cosmos to conform with our behavior, we would have to refashion our behavior to conform with the cosmos.

But how would we know what the cosmos expected of us? To

begin with, we would need to acknowledge the gap that will forever exist between the world we experience and the experience of the world. The gap between the two, the great, fathomless void that separates us from everything beyond our reach, is not where all of life ends, but where all of life begins. It is the place where security is nonexistent, where all things are vulnerable, where there are no hierarchies, no pecking orders, only relationships and mutual dependencies. This is the world we are frightened of, the world we block from our consciousness. We much prefer the world of self-containment. Like a forlorn creature cast adrift on a thin sheet of ice, we float on top of a sea with no boundaries, convinced that as long as we remain dry we will remain invulnerable. We can't afford to believe that the ice too is water and will melt back into the depths. Our cosmologies are like thin vessels of ice. They are made of the real world, but they do not define it. They exist for a moment of time but are eventually submerged, and each time they carry us down with them. If only we could learn to swim, then we would never need to flail around for yet another icy perch to climb onto. But to swim we would need to renounce our drive for self-containment. We would have to make a conscious decision to exchange security for vulnerability and give up our power over the future.

To choose participation when we could control. To acknowledge relationship when we could dominate. Few among us would be willing to make such an exchange. It is far easier to live with our own self-deception, to continue the pretense that the world is made for us rather than to acknowledge that we are made for the world. If the truth be known, we really don't want to belong to the world. We want the world to belong to us. And as long as that is the case, we will continue to seek new ways to organize the world into ourselves and will continue to justify each new form of appropriation with the suitable cosmological reformulation. We will never reject any effort to extend our control over the forces of nature as long as such efforts guarantee us a measure of security against the vagaries of the future. As in the past, responsibility will never be a question, because in order to be responsible to something, one has to feel a sense of obligation and indebtedness, and that requires an acknowledgment of relationship and mutual dependency. As long as humanity is preoccupied by the drive for self-containment, everything else will exist as material for manipulation. It can be no other way. If humanity believes itself to be self-contained, it cannot believe itself to be indebted at the same time, despite attempts over the millennia to unite the two. Either we are not in need of anything outside of ourselves or we are. If we're not in need of anything outside ourselves, then we can do pretty much as we please. If, on the other hand, we are in need of things outside ourselves, then we are indebted and must take into account more than our own desires when we act.

The great void that we are so terrified of is filled with indebtedness. As long as we ignore the cosmic ledger, we can continue to organize the world exclusively for ourselves without ever entertaining so much as a smidgen of doubt as to the rightness of our behavior. The moment we introduce indebtedness, the entire picture changes. Suddenly we are surrounded by a multitude of relationships, all of which have contributed to our perpetuation. We live by the grace of sacrifice. Every amplification of our being owes its existence to some diminution somewhere else. In an ultimate sense nothing that we claim as ours belongs to us, not even our fiber and sinew. Everything about us has been borrowed. We have been lent by nature. It has given over to us parts of itself, thus precluding their use for an infinite number of other things. The cosmos owes nothing to us. We owe everything to the cosmos.

By our very existence, then, we are accountable. The great question yet to be addressed by the human race is whether or not we will choose to acknowledge the debt of our existence. Acknowledgment requires that for the first time in history we act not only for ourselves but for everything else in the universe we have a relationship with. To act in such a manner as to represent the interests of the cosmos: this is the meaning of the word "responsibility," a word that cosmologists thus far have attempted to ignore. The interests of the cosmos are no different from ours. We seek to be, to perpetuate, to continue to exist. It is reasonable to assume that so too does everything else. But it is also true that perpetuation requires extinction, that life requires death. How then do we best represent the interests of the cosmos? By paying back to the extent to which we received. By sacrificing to the cosmos the measure of the sacrifices the cosmos made for us. . . .

Sacrifice requires, above all, that humanity give up a degree of control over its future. That's because human security is always exacted from nature's sacrifice. We take from nature to secure ourselves. The human sojourn has been a relentless drive to gain greater control over the future in order to secure our own perpetuation. Nature has given up more and more of itself to us so that we can secure more and more of our future. There is only one way to pay back such a debt: to sacrifice some of our security in order for nature to secure itself.

How different this pure form of sacrifice is from the bastardized version we so frequently employ as a token to expiate our guilt. Pure sacrifice, collectively expressed and generationally sustained, has never yet occurred anywhere on earth.

Sacrificing our security requires a wholesale transformation in the way we approach the future. To begin with, our concept of knowledge will have to be redefined. Increasingly, Western men and women have sought knowledge in order to control their environment. We pursue knowledge so that we may better predict what lies ahead. The goal of knowledge has been foresight, and Western civilization has used foresight to tighten its grip on the becoming process. We need to pursue a different knowledge path, a path whose goal is to foresee how better to participate with rather than to dominate nature. To better understand the why of things as opposed to the how of things. This is the knowledge of relationship and is in strong contrast to the knowledge of usurpation that so obsesses the modern mind.

Today we are well versed in how to pursue technological knowledge but virtually untutored when it comes to pursuing empathetic knowledge. Technological knowledge gives us foresight so that we can better appropriate the life around us. Empathetic knowledge gives us foresight so that we can better cooperate with the community of life. With technological foresight, security comes in exercising power over nature. With empathetic foresight, security comes from belonging to a community.

To end our long, self-imposed exile; to rejoin the community of life. This is the task before us. It will require that we renounce our drive for sovereignty over everything that lives; that we restore the rest of creation to a place of dignity and respect. The resacralization of nature stands before us as the great mission of the coming age.

Two futures beckon us. We can choose to engineer the life of the planet, creating a second nature in our image, or we can choose to participate with the rest of the living kingdom. Two futures, two choices. An engineering approach to the age of biology or an ecological approach. The battle between bioengineering and ecology is a battle of values. Our choice, in the final analysis, depends on what we value most in life. If it is physical security, perpetuation at all costs, that we value most, then technological mastery over the becoming process is an appropriate choice. But the ultimate and final power to simulate life, to imitate nature, to fabricate the becoming process brings with it a price far greater than any humanity has ever had to contend with. By choosing the power of authorship, humanity gives up, once and for all, the most precious gift of all, companionship.

As bioengineering technology winds its way through the many passageways of life, stripping one living thing after another of its identity, replacing the original creations with technologically designed replicas, the world gradually becomes a lonelier place. From a world teeming with life, a world spontaneous, unpredictable, dynamic, rhapsodizing, we descend to a world stocked with living gadgets and devices, a world running smoothly, effortlessly, quietly, without feeling. In the end, it is companionship we give up, the companionship with other life that is at once both indescribable and essential, and without which existence becomes a meaningless exercise.

There could be no lonelier place than a biologically engineered world. That's why even if only one living creature were left unscathed in a world brimming over with biological facsimiles, we would reach out to it, embrace it, touch it, marvel at it, with a peak of emotion that all the replicas together could not possibly hope to tap in us. For we experience something special with that creature that can never be experienced with the products of our own technological handiwork: a bond, steeped in the mystery of a common origin to which we are both beholden and to which there is no way of ever giving proper thanks.

If, on the other hand, it is companionship and belonging we value most, then an ecological approach to the coming age of biology is the appropriate course. But even here there is a price to be paid. Companionship requires sacrifice, the willingness to risk one's physical security in order to protect the interests of the community one is a part of. Companionship requires participation, sharing, and above all vulnerability. It is the price one pays to belong, to be a member in good standing in the community of life.

Our choices, then, are not easy ones. Giving up bioengineering means sacrificing a measure of control over the future. Compromising our drive for total mastery over what lies ahead. Making ourselves more vulnerable so that the rest of existence can become more secure. Choosing to serve and nurture even though we have it in our power to dominate and extract. These fly in the face of the human experience to date. When it comes to securing our future, we have never once flinched from a total unswerving commitment. Over and over again we have fashioned new, more ingenious ways to organize our future security. Each time nature sacrificed so that we might triumph. And each time we constructed a new image of the universe that glorified and sanctified our new extractive relationship with the world around us.

Not once in the long history of Western civilization have we ever said no to securing our own future. We have organized ourselves and the world, then reorganized, then reorganized again. Over and over, we have reworked, refashioned, and revised. Each time we were looking for something of overwhelming import, something we hoped would not be denied us—our immortality. We have spent the world hoping to buy our permanence, only to discover that our everlasting presence is to be found in what remains untouched, not in what we use up. This is the lesson, if there is one, in the long journey of self-deception that Western men and women have taken over the millennia.

We have advanced on the world, swept across the great landscapes, leaving our mark in granite and stone, iron and steel, to let whatever follows know that we're still here, that our presence remains. But these things we organize are not of life but of death. Our monuments, our edifices, our inventions are things that have passed. Everything we leave can only be a corpse; some remains of what was once a possibility and is now an aftermath, a finality. What is alive is in the next moment; what is dead is the last moment. Immortality can exist only in what is potential, not in what has already been. What lives forever is the potential we left behind, the great possibilities we chose not to squander. For millennia we left a legacy of death, when in fact the only living legacy that we can ever leave is the endowment we never touched. Our presence in the future will not be felt by what we spent but by what we left unspent.

We gain immortality through our sacrifices. It is by giving something back, by leaving something unspent, by going without, that we live on. This is our present to the future. It is the only real legacy. By leaving behind all of the unspent possibilities, we bequeath the finest endowment of all, the gift of life for future generations to enjoy.

This does not mean that our own lives should barely be lived; that the best course of action is virtual inaction so as not to use any more of the resources than absolutely necessary to merely survive. It does mean that we need to continually ask ourselves how much is enough, and be willing to discipline our appetites so that they remain within the bonds dictated by a sense of fair regard for every other living thing. It means that each time we consider a course of action, we ask ourselves how our decision will affect the well-being of the rest of the living kingdom today as well as that of future generations.

The skeptic might inquire as to why we should show a fair regard for every other living thing. The answer is to be found in how we come to terms with our own existence. What is the purpose of life? Why are we here? When confronted with our own existence, two choices present themselves. To accept life as a gift to be enjoyed or as an obstacle to overcome. If we experience life as a gift, we give thanks. Giving thanks means sharing our good fortune by helping to extend the gift of life to the rest of posterity. Indeed, if wisdom exists at all, it resides in the knowledge that life can be truly enjoyed only if it is generously shared and extended. If, however, we experience life as an obstacle to overcome, then we will be relentless in our search for ways to defeat its most essential attribute, its temporary nature, its limited duration. We will devour the life around us in order to extend our own. We will exhaust the very reservoirs of life from which the future is secured, all in an effort to secure our own future in perpetuity.

Up to now we have allowed nothing to come in the way of our efforts to secure our future. Over the centuries, we have constructed countless cosmologies to lend an air of legitimacy to our ceaseless drive for self-perpetuation at all costs. We have deceived ourselves into believing that our interests were in accord with the interests of the universe, when in fact it was only our limited needs that were being projected onto the cosmos.

Sacrificing a measure of our own future security in order to represent the interests of the rest of the cosmos is the most difficult request that we can ever make of ourselves. It is fitting and perhaps more than a bit ironic that only now is the human race even entertaining such a question for the first time. Now that we have it within our power to refashion all of nature in order to secure our future, we need to ponder whether such is our right.

Can any of us imagine making such a sacrifice, giving up a measure of control over our own future? Can any of us imagine saying no to all the great benefits that the bioengineering of life will bring to bear? Can any of us, for that matter, entertain even for a moment the prospect of saying no to the age of biotechnology? If we cannot even entertain the question, then we already know the answer. Our future is secured. The cosmos wails.

NOTES

<hr>

I FROM ALCHEMY TO ALGENY:
A New Metaphor for the Coming Age

1. Lewis Mumford, *Technics and Human Development: The Myth of the Machine* (New York: Harcourt, Brace & World, 1966), vol. I, page 124.

2. Wertime and James D. Muhly, *The Coming of the Age of Iron* (New Haven: Yale University Press, 1980), page 9.

3. Quoted in Erik Eckholm, "Disappearing Species: The Social Challenge," *Worldwatch Paper 22* (Washington, D.C.: Worldwatch Institute, June 1978), page 6.

4. Lawrence Lessing, "Into the Core of Life Itself," *Fortune* (March 1966), page 150.

5. Ibid., page 152.

6. Ibid.

7. Joel N. Shurkin, "Yet Another Step in the Complex Probe of the Genetic Code," *Philadelphia Inquirer* (May 1, 1977).

8. "Fetus for Sale," *Newsweek* (June 1, 1970), page 86.

9. Graham Chedd, "Danielli the Prophet," *New Scientist*, 49 (January 21, 1971), page 124.

10. D. S. Halacy, Jr., *The Genetic Revolution* (New York: Harper & Row, 1974), page 185.

11. *DNA Recombinant Molecule Research,* Supplemental Report II, Report prepared for the Subcommittee on Science, Research and Technology of the Committee on Science and Technology, House of Representatives, December 1976, page 16.

12. Jonathan Beckwith, "Gene Expression in Bacteria and Some Concerns About the Misuse of Science," *Bacteriological Review,* 34 (1970), page 224.

13. Jonathan Beckwith, "Recombinant DNA: Does the Fault Lie Within Our Genes?" Paper presented to the National Academy of Sciences Forum on Recombinant DNA, Washington D.C., March 7–9, 1977.

14. "Counting Up the Genes," *The New York Times* (April 24, 1977), Section 4, page 6.

15. "A Working Synthetic Gene," *Medical World News* (September 20, 1976), page 7.

16. Carl R. Merril, et al., "Bacterial versus Gene Expression in Human Cells," *Nature,* 233 (1971), pages 398–400.

17. Personal interview with Dr. Ethan Signer, April 1977.

18. "Retailoring the Tailor," *1976 Encyclopedia Britannica, Book of the Year,* Special Supplement, page iv.

19. *Impacts of Applied Genetics* (Washington, D.C.: U.S. Government Printing Office, 1981), page 8.

20. Sharon and Kathleen McAuliffe, *Life for Sale* (New York: Coward, McCann & Geoghegan, 1981), page 28.

21. Ibid., page 11.

22. Ibid., page 26.

23. Thomas O'Toole, "In the Lab: Bugs to Grow Wheat, Eat Metal," *The Washington Post* (June 18, 1980), page A1.

24. Quoted in Victor Cohn, "Biologists Report Transfer of Gene from Rabbit to Mouse to Offspring," *The Washington Post* (September 8, 1981), page A12.

25. (Ithaca: Cornell University Press, 1981), page 92.

26. Ibid., page 88.

27. Quoted in Titus Burckhardt, *Alchemy,* trans. William Stoddart (London: Stuart & Watkins, 1967), page 25.

28. Quoted in Edward B. Fiske, "Computers Alter Life of Pupils and Teachers," *The New York Times* (April 4, 1982), page 1.

29. Ibid.

30. Ibid.

31. Joseph A. Menosky, "Cheap, Fast Designer Genes," *The Washington Post* (September 6, 1981), page C1.

32. Quoted in Kathleen McAuliffe, "Biochip Revolution," *Omni* (March 3, 1982), page 55.

33. Ibid.

34. *Life for Sale,* page 219.

II DECIDING WHAT'S NATURAL:
The Ultimate Intellectual Deception

1. Quoted in Loren Eiseley, *The Firmament of Time* (New York: Atheneum, 1960), page 44.

2. Michael Mulkay, *Science and Sociology of Knowledge* (London: George Allen & Unwin, 1979), page 1.

3. Ibid., page 2.

4. *The Foundation of Primitive Thought* (Oxford: Clarendon Press, 1979), page 480.

5. Samuel Noah Kramer, *The Sumerians* (Chicago: University of Chicago Press, 1963), page 113.

6. Ibid., page 114.

7. William Irwin Thompson, *The Time Falling Bodies Take to Light* (New York: St. Martin's Press, 1981), page 162.

8. Ibid.

9. Quoted in L. von Bertalanffy, *Robots, Men and Minds* (New York: Braziller, 1967), page 91.

10. *Traces on the Rhodian Shore* (Berkeley: University of California Press, 1967), page 230.

11. Ibid.

12. Quoted in Arthur O. Lovejoy, *The Great Chain of Being* (Cambridge: Harvard University Press, 1936), page 86.

13. *Europe in the Middle Ages,* 2nd ed. (New York: Harcourt, Brace & World, 1966), page 300.

14. Lovejoy, pages 77–78.

15. Ibid., page 79.

16. Ibid.

17. *Beyond the Pleasure Principle,* trans. by J. Strachey (New York: W. W. Norton, 1961), page 32.

18. Ernest Becker, *Escape from Evil* (New York: Free Press, 1975), page 3.

19. Ibid.

20. Ibid.

21. Ibid., page 63.

22. Quoted in ibid., page 136.

23. Ibid., pages 136–37.

24. *Crowds and Power,* trans. by Carol Stewart (London: Gollancz, 1962), page 448.

25. *Nature's Economy: The Roots of Ecology* (Garden City: Anchor Books, 1979), page 96.

26. *The World of Primitive Man* (New York: Grove Press, 1953), page 3.

27. *The Psychology of Science* (New York: Harper & Row, 1966), page 139.

28. Ibid.

29. *The Technological Society,* trans. by John Wilkinson (New York: Knopf, 1964), pages 141–42.

30. *Social Origins,* foreword by Lord Raglan (London: Watts, 1954), page 35.

31. Quoted in Charles Darwin, *The Origin of Species* and *The Descent of Man* (New York: Modern Library, 1936), page 492.

32. Becker, pages 22–23.

33. Ibid., page 149.

III DARWIN'S VISION:
A Reflection of the Industrial State of Mind

1. *Beyond Psychology* (New York: Dover, 1941), pages 32–33.

2. *Science, Ideology, and World View* (Berkeley: University of California Press, 1981), page 124.

3. Ibid., page 7.

4. Ibid.

5. "Social Factors in the Origin of Darwinism," *Quarterly Review of Biology,* 13 (1938), page 325.

6. Greene, page 7.

7. Robert Young, "Man's Place in Nature," *Changing Perspectives in the History of Science,* eds. Mikulas Teich and Robert Young (Boston: R. Reidel Publishing, 1975), page 384.

8. Quoted in Richard D. Altick, *Victorian People and Ideas* (New York: W. W. Norton, 1973), page 11.

9. Ibid.

10. Jacquetta and Christopher Hawkes, "Land and People," *The Character of England,* ed. Ernest Barker (Oxford: Clarendon Press, 1947), page 27.

11. Altick, page 41.

12. *The Growth of Cities in the Nineteenth Century* (New York: Macmillan, 1899), page 1.

13. "History, Philosophy and Sociology of Biology: A Family Romance," *Studies in the History and Philosophy of Science,* 2 (1980), page 3.

14. Ibid.

15. Altick, page 43.

16. Ibid., page 45.

17. Ibid.

18. Ibid., page 81.

19. E. J. Hobsbawm, *The Age of Capital, 1848–1875* (New York: Scribner's, 1975), page 31.

20. Ibid., page 30.

21. Martin Wiener, page 12.

22. *Charles Darwin: A Portrait* (New Haven: Yale University Press, 1938), page 334.

23. Ibid., page 335.

24. Ibid.

25. Ibid.

26. *Autobiography of Charles Darwin: 1809–1882,* ed. Nora Barlow (New York: W. W. Norton, 1958), page 23.

27. Ibid.

28. Charles Darwin, *Life & Letters,* ed. Francis Darwin (New York: Appleton, 1887), vol. 1, page 290.

29. West, page 334.

30. *Darwin and the Darwinian Revolution* (New York: W. W. Norton, 1959), page 128.

31. Darwin, *Life and Letters,* vol. 1, page 243.

32. Quoted in Himmelfarb, pages 129–30.

33. Ibid., page 136.

34. Ibid.

35. Ibid., page 137.

36. Worster, page 13.

37. Ibid.

38. Sandow, page 321.

39. Barlow, page 119.

40. Sandow, page 321.

41. Peter J. Vorzimmer, "An Early Darwin Manuscript: The 'Outline and Draft of 1839,' *Journal of the History of Biology,* 8 (1975), page 215.

42. Charles Darwin, *The Variation of Animals and Plants Under Domestication* (New York: Appleton, 1896), vol. I, page 447.

43. Mulkay, page 102.

44. Ibid., page 104.

45. Charles Darwin, *The Origin of Species* (New York: Watts, 1929), page 72.

46. Barlow, page 120.

47. Worster, page 150.

48. Mulkay, page 105.

49. Worster, page 151.

50. *Religion and Science* (Oxford: Oxford University Press, 1935), pages 72–73.

51. "Darwin and the Political Economists," *Journal of the History of Biology,* 13 (1980), page 200.

52. Mulkay, pages 105–106.

53. Darwin, *Life and Letters,* vol. 1, page 531.

54. "Outline of an Historical View of the Progress of the Human Mind," *Main Currents in Modern Political Thought* (New York: Holt, Rinehart & Winston, 1950), page 132.

55. "Darwin's Mistake," *Harper's* (February 1976), page 71.

56. Darwin, *The Origin of Species,* page 209.

57. Worster, page 149.

58. *Herman Melville,* ed. R. W. B. Lewis (New York: Dell, 1962), page 126.

59. Worster, page 124.

60. Charles Darwin, *The Voyage of the Beagle,* ed. Leonard Engel (Garden City, N.Y.: Doubleday, 1962), page 375.

61. Ibid.

62. Barry Gale, "Darwin and the Struggle for Existence: A Study in the Extrascientific Origins of Scientific Ideas," *Isis,* 63 (1972), page 343.

63. Barlow, page 92.

64. Schweber, page 198.

65. Adam Smith, *An Inquiry into the Nature and Causes of the Wealth of Nations* (Oxford: Clarendon Press, 1976), page 13.

66. Quoted in Schweber, page 274.

67. Quoted in ibid., page 251.

68. Quoted in ibid.

69. Quoted in ibid.

70. Quoted in ibid., pages 255–56.

71. Quoted in ibid., page 252.

72. Ibid., page 256.

73. Ibid., page 197.

74. Ibid.

75. Barlow, page 120.

76. Charles Darwin and A. R. Wallace, *Evolution by Natural Selection* (Cambridge, England: Cambridge University Press, 1958), pages 264–67.

77. *Charles Darwin's Natural Selection,* ed. R. C. Stauffer (Cambridge, England: Cambridge University Press, 1975), page 233.

78. Ibid., page 228.

79. Page 375.

80. Darwin, *The Origin of Species,* page 87.

81. Worster, page 158.

82. Ibid., page 162.

83. Ibid.

84. Ibid., page 161.

85. Ibid.

86. Quoted in Howard E. Gruber and Paul H. Barrett, *Darwin on Man: A Psychological Study of Scientific Creativity* (New York: E. P. Dutton, 1974), page 459. (Notebook E., *Transmutation of Species.*)

87. Darwin, *Life and Letters,* vol. I, page 316.

88. Himmelfarb, page 425.

89. Ibid.

90. Charles Darwin, *The Descent of Man and Selection in Relation to Sex* (New York: Appleton, 1896), page 134.

91. *The Principles of Political Economy,* 2nd ed. (London: Longman, Rees, Orme, Brown & Green, 1830), page 537.

92. Ibid., page 149.

93. Darwin and Wallace, page 119.

94. "Malthus, Darwin and the Concept of Struggle," *Journal of the History of Ideas,* 37 (1976), page 645.

95. *Social Statics* (London: J. Chapman, 1851), pages 322–23.

96. Richard Hofstadter, *Social Darwinism in American Thought*, rev. ed. (Boston: Beacon Press, 1955), page 7.

97. Ibid., pages 6–7.

98. *Darwin and the Problem of Creation* (Chicago: University of Chicago Press, 1979), page 54.

99. Quoted in Loren Eiseley, *The Firmament of Time* (New York: Atheneum, 1960), page 29.

100. Quoted in Himmelfarb, page 337.

101. Worster, page 40.

102. "The Ideal World View," *The Schumacher Lectures* (New York: Harper and Row, 1981), page 97.

103. Quoted in Himmelfarb, page 348.

104. Ibid., page 347.

105. Darwin, *The Origin of Species*, page 72.

106. Spencer, pages 79–80.

107. Schweber, page 277

108. Worster, page 52.

109. West, page 324.

110. Ibid.

111. Karl Marx and Friedrich Engels, *Correspondence*, trans. & ed. Dona Torr (London: International Publishers, 1934), page 125.

112. Himmelfarb, page 421.

113. Karl Marx, *Selected Works*, ed. E. P. Dutt (London: Lawrence & Wishart Ltd., 1943), vol. I, page 16.

114. Himmelfarb, page 423.

115. Marx and Engels, *Selected Correspondence*, trans. by I. Lasker, page 302.

116. *The Decline of the West* (New York: Knopf, 1939), page 373.

117. Himmelfarb, page 416.

IV THE DARWINIAN SUNSET:
The Passing of a Paradigm

1. Julian Huxley, "At Random—A Television Preview," *Evolution after Darwin*, ed. Sol Tax (Chicago: University of Chicago Press, 1960), vol. I, page 42.

2. Ibid., page 41.

3. *Nature and Man's Fate* (New York: Mentor Books, 1961), page 216.

4. Speech delivered at the American Museum of Natural History, New York, N.Y., November 5, 1981.

5. *The Flight from Women* (New York: Farrar, Straus & Giroux, 1965), page 290.

6. Ibid.

7. Everett Claire Olson, "The Evolution of Life," *Evolution after Darwin*, vol. I, page 523.

8. Ibid.

9. John Davy, "What If Darwin Were Wrong?," *The Washington Post* (August 30, 1981), page C1.

10. Ibid.

11. (New York: Pergamon Press, 1960), page vii.

12. John Tyler Bonner, "Implications of Evolution," *American Scientist* 49 (1961), page 240.

13. "Darwinian or 'Oriented' Evolution?," *Evolution*, 29 (1975), page 376.

14. *Evolution of Living Organisms* (New York: Academic Press, 1977), page 202.

15. Ibid., page 6.

16. Introduction to *The Origin of Species* (London: J. M. Dent and Sons, 1971), page xi.

17. "Notes on the Nature of Science by a Biologist," *Notes on the Nature of Science* (New York: Harcourt, Brace & World, 1962), page 9.

18. "On Methods of Evolutionary Biology and Anthropology," *American Scientist*, 45 (1957), page 388.

19. Ibid.

20. *Life: Its Nature, Origin and Development* (Edinburgh: Oliver and Boyd, 1961), page 33.

21. Quoted in Robert T. Clark and James D. Bales, *Why Scientists Accept Evolution* (Grand Rapids, Mich.: Baker Book House, 1966), page 95.

22. (New York: E. P. Dutton, 1956.)

23. *Man Real and Ideal* (New York: Scribner's, 1943), page 147.

24. Ibid.

25. "Evolution," *Scientific American*, 239 (September 1978), page 52.

26. Ibid.

27. Ibid., pages 52–53.

28. *Evolution, the Modern Synthesis* (London: Allen and Unwin, 1942), page 28.

29. Roger Lewin, "Evolutionary Theory under Fire," *Science*, 210 (1981), page 883.

30. "Is a New and General Theory of Evolution Emerging," *Paleobiology*, 6 (1980), page 120.

31. Grassé, page 4.

32. Ibid., page 31.

33. Ibid.

34. Darwin, *The Origin of Species*, page 239.

35. Quoted in Gary E. Parker, *Creation: The Facts of Life* (San Diego: Creation Life Publishers, 1980), pages 94–95.

36. "Early Cambrian Marine Fauna," *Science*, 128 (1958), page 7.

37. *The Meaning of Evolution* (New Haven: Yale University Press, 1949), page 18.

38. *The Fishes* (New York: Life Nature Library, Time-Life, Inc., 1964), page 60.

39. *Vertebrate Paleontology*, 3rd. ed. (Chicago: University of Chicago Press, 1966), page 15.

40. "Proceedings," *Linnaean Society of London*, 177 (1966), page 8.

41. *The Fossils Say No!* (San Diego: Creation Life Publishers, 1978), pages 74–75.

42. *Biology and Comparative Physiology of Birds*, ed. A. J. Marshall (New York: Academic Press, 1960), vol. I, page 1.

43. "Bone Bonanza: Early Bird and Mastodon," *Science News*, 112 (September 24, 1977), page 198.

44. PBS Television Show, "Did Darwin Get It Wrong?" Originally broadcast on November 1, 1981. WGBH Transcripts, 125 Western Ave., Boston, Mass.

45. "Evolution, as Viewed by One Geneticist," *American Scientist*, 40 (1952), page 97.

46. "Paleontology and Evolutionary Theory," *Evolution*, 28 (1974), page 467.

47. Stephen Jay Gould and Niles Eldredge, "Punctuated Equilibria: The Tempo and Mode of Evolution Reconsidered," *Paleobiology*, 3 (1977), page 115.

48. Ibid.

49. Quoted in Norman Macbeth, *Darwin Retried* (Boston: Gambit Inc., 1971), pages 35–36.

50. "The Reply: Letter from Birnam Wood," *Yale Review*, 56 (1967), page 636.

51. Quoted in Macbeth, page 36.

52. Grassé, pages 87–88.

53. Ibid.

54. Ibid., page 202.

55. *Introduction to Quantitative Genetics* (New York: Ronald Press, 1960), page 186.

56. Quoted in Parker, page 76.

57. R. L. Wysong, *Creation–Evolution Controversy* (Midland, Mich.: Inquiry Press, 1976), page 274.

58. *Animal Species and Evolution* (Cambridge: Harvard University Press, 1963), pages 285–86.

59. Ibid., page 290.

60. Wysong, page 274.

61. Himmelfarb, page 446.

62. Quoted in David L. Willis, "Creation and/or Evolution," *Origins and Change* (Elgin, Ill.: A. S. Air, 1978), page 9.

63. Grassé, page 88.

64. Ibid.

65. Himmelfarb, page 341.

66. Ibid., pages 341–42.

67. Ibid., page 342.

68. Stephen Jay Gould, "The Return of Hopeful Monsters," *Natural History,* 86 (June–July 1977), page 24.

69. Macbeth, pages 99–100.

70. Himmelfarb, pages 337–38.

71. Wysong, page 306.

72. Darwin, *Life and Letters,* vol. II, page 67.

73. Darwin, *The Origin of Species,* page 160.

74. *Natural Selection and Tropical Nature* (London: Macmillan, 1895), page 202.

75. Quoted in Loren Eiseley, *Darwin's Century* (New York: Doubleday, 1958), page 311.

76. Macbeth, page 103.

77. "The Evolutionary Biology of Constraint," *Daedalus,* 29 (1980), page 46.

78. John Arthur Thomson and Patrick Geddes, *Life: Outlines of General Biology* (London: Williams & Norgate, 1931), vol. II, page 1317.

79. *Evolution in Action* (New York: Mentor, 1957), page 34.

80. Quoted in Bethell, page 72.

81. Himmelfarb, page 316.

82. *The Strategy of the Genes* (London: Allen & Unwin, 1957), pages 64–65.

83. Book review of *Der gerechtfertigte Haeckel and Nomogenesis, or Evolution Determined by Law,* in *Science,* 164 (1969), page 684.

84. *Embryos and Ancestors* (New York: Oxford University Press, 1954), page 6.

85. Ibid.

86. "Evolution," *New Scientist,* 49 (1971), page 35.

87. Wysong, page 398.

88. *The Origin of Life* (New York: Reinhold, 1968), page 27.

89. "The Folly of Probability," *The Origins of Prebiological Systems and Their Molecular Matrices,* ed. S. W. Fox (New York: Academic Press, 1965), page 41.

90. "Production of Some Organic Compounds under Possible Primitive Conditions," *Journal of the American Chemical Society,* 77 (1955), page 2351.

91. Wysong, page 212.

92. A. E. Wilder-Smith, *The Natural Sciences Know Nothing of Evolution* (San Diego: Master Books, 1981), page 14.

93. Ibid., page 19.

94. Ibid., page 20.

95. Quoted in Wysong, pages 75–76.

96. Kerkut, page 150.

97. Quoted in Walter Sullivan, "Creation Debate Is Not Limited to Arkansas Trial," *The New York Times* (December 27, 1981), page 48.

98. Wysong, page 195.

99. Ibid., page 104.

100. Simpson, *The Meaning of Evolution,* pages 15–16.

101. Hadd, page 31.

102. Ibid.

103. Parker, page 35.

104. "The Evolutionary Paradox and Biological Stability," *Molecular Evolution: Prebiological and Biological,* eds. D. L. Rohlfing and A. I. Oparin (New York: Plenum Press, 1972), page 111.

105. Ibid.

106. "Algorithms and the Neo-Darwinian Theory of Evolution," *Mathemati-*

cal Challenges to the Neo-Darwinian Interpretation of Evolution, eds. P. Moorhead and M. Kaplan (Philadelphia: Wister Institute Press, 1967), pages 74–75.

107. Murray Eden, "Inadequacies of Neo-Darwinian Evolution as a Scientific Theory," *Mathematical Challenges to the Neo-Darwinian Interpretation of Evolution,* page 109.

108. Schutzenberger, page 75.

109. *Janus: A Summing Up* (New York: Random House, 1978), page 192.

110. Ibid.

111. Quoted in Koestler, page 179.

V RETHINKING AND REMAKING LIFE:
A New "Temporal" Theory of Nature
Custom-made for the Biotechnical Age

1. *Epigenetics* (London: John Wiley & Sons, n.d.), page 404.

2. M. W. Ho and P. T. Saunders, "Beyond Neo-Darwinism—An Epigenetic Approach to Evolution," *Journal of Theoretical Biology,* 78 (1979), page 579.

3. *The Evolution of an Evolutionist* (Edinburgh: Edinburgh University Press, 1975), page vi.

4. C. H. Waddington, "A Catastrophe Theory of Evolution," *Annals of the New York Academy of Science,* 231 (1974), page 36.

5. Ho and Saunders, page 580.

6. Robert E. Monro, "Interpreting Molecular Biology," *Beyond Chance and Necessity,* ed. John Lewis (London: Garnstone Press, 1974), page 108.

7. (January 11, 1982), page 39.

8. Ibid.

9. Quoted in ibid.

10. Ibid., page 40.

11. "Neo-Darwinism," *Theoria to Theory,* 13 (1979), page 201.

12. (New Haven: Yale University Press, 1976), page 60.

13. *The Science of Life* (New York: Futura Publishing, 1973), page 19.

14. Ibid., page 42.

15. Ibid.

16. Ibid., page 19.

17. Ibid., page 45.

18. Ibid., page 47.

19. Ibid., page 61.

20. Ibid.

21. "On Morphogenetic Fields," *Theoria to Theory*, 13 (1979), page 111.

22. Ibid., page 112.

23. Ibid.

24. Ibid.

25. Ibid., page 113.

26. Ibid., pages 113–14.

27. *Blueprint for Immortality* (London: Neville Spearman, 1972), page 30.

28. Ibid.

29. Ibid., page 33.

30. Ibid., pages 43–44.

31. Ibid., page 71.

32. Ibid., page 107

33. Ibid.

34. "The Rhythmic Nature of Animals and Plants," *American Scientist*, 47 (1959), page 159.

35. Ibid., page 161.

36. "Circadian Rhythms in Man," *Science*, 148 (1965), page 1431.

37. "The Cell as a Resonating System," *Towards a Theoretical Biology*, ed. C. H. Waddington (Chicago: Aldine, 1969), vol. 2, page 198.

38. J. T. Fraser, *Of Time, Passion, and Knowledge: Reflections on the Strategy of Existence* (New York: George Braziller, 1975), page 186.

39. Brown, page 166.

40. Quoted in *Bergson and the Evolution of Physics*, ed. P. A. Y. Gunter (Knoxville, Tenn.: University of Tennessee Press, 1977), page 63.

41. "Temporal Order and Spatial Order: Their Differences and Relations," *Mind in Nature*, eds. John Cobb and David Griffin (Washington, D.C.: University Press of America, 1977), page 80.

42. Quoted in John Dewey, "Metaphysical Issues: The Influence of Darwinism," *The Problem of Evolution*, eds. John Deely and Raymond Nogar (New York: Appleton-Century-Crofts, 1973), page 259.

43. *The Wonders of Life* (New York: Harper & Brothers, 1905), page 80.

44. Worster, page 209.

45. *The Idea of Nature* (Oxford: Clarendon Press, 1945), page 146.

46. Quoted in Alfred North Whitehead, *The Principles of Natural Knowledge,* 2nd ed. (Cambridge, England: Cambridge University Press, 1925), page 54.

47. Collingwood, page 146.

48. *Nature and Life* (New York: Greenwood Press, 1968), pages 20–22.

49. Ibid., page 12.

50. Ibid., page 15.

51. Ibid., page 27.

52. Whitehead, *The Principles of Natural Knowledge,* page 6.

53. Capek, page 51.

54. *The Idea of Nature* (Oxford: Clarendon Press, 1945), page 24.

55. Fraser, page 80.

56. Charles Hartshorne, "Physics and Psychics: The Place of Mind in Nature," *Mind in Nature,* page 92.

57. Fraser, page 442.

58. Ibid., pages 224–25.

59. Ibid., page 228.

60. Stephen Jay Gould and Niles Eldredge, "Punctuated Equilibria: The Tempo and Mode of Evolution Reconsidered," *Paleobiology,* 3 (1977), page 115.

61. Quoted in Edward B. Fiske, "Computers Alter Life of Pupils and Teachers," *The New York Times* (April 4, 1982).

62. "Computer Age Began with Marvel of 1951," *The Washington Star* (June 10, 1981), page C8.

63. Alvin Toffler, *The Third Wave* (New York: Bantam, 1980), page 140.

64. Ibid., page 170.

65. Carl Mitcham, "Philosophy of Technology," *Science, Technology, Medicine,* ed. Paul T. Durbin (New York: Free Press, 1980), page 316.

66. *The Human Use of Human Beings* (New York: Avon Books, 1954), pages 26–27.

67. Ibid., page 35.

68. A. Rosenblueth, N. Wiener, and J. Bigelow, "Behavior, Purpose and Teleology," *Philosophy of Science,* 10 (1943), page 18.

69. Wiener, page 278.

70. Ibid., page 25.

71. (Berkeley: University of California Press, 1972), page 194.

72. Ibid.

73. Roy A. Rappaport, *Ecology, Meaning, and Religion* (Richmond, Calif.: North Atlantic Books, 1979), page 169.

74. Toffler, page 185.

75. Ibid.

76. *The Social Impact of Cybernetics*, ed. Charles R. Dechert (New York: Simon and Schuster, 1966), pages 18–19.

77. "Cybernetics and the Problems of Social Reorganization," *The Social Impact of Cybernetics*, page 59.

78. *Understanding Media: The Extensions of Man* (New York: New American Library, 1964), pages 302–303.

79. Ibid., page 53.

80. Ibid., page 225.

81. Ibid., page 103.

82. (Cambridge: Cambridge University Press, 1980), page 303.

83. Ibid.

84. *The Understanding of Nature* (Dordrecht: D. Reidel, 1974), page 68.

85. "The Brain as an Engineering Problem," *Current Problems in Animal Behavior,* eds. W. H. Thorpe and O. L. Zangwill (Cambridge, England: Cambridge University Press, 1961), page 307.

86. "The Frontiers of Biology," *Mind in Nature,* eds. John Cobb and David Griffin (Washington, D.C.: University Press of America, 1977), page 3.

87. C. H. Waddington, "Whitehead and Modern Science," *Mind in Nature,* page 145.

88. Ibid.

89. "The Frontiers of Biology," page 6.

90. *Cybernetics and the Philosophy of Mind* (Highlands, N.J.: Humanities Press, 1976), page xi.

91. Grassé, page 223.

92. Ibid., page 224.

93. Ibid.

94. Ibid., page 225.

95. Ibid., page 226.

96. Ibid., page 225.

97. John J. Ford, "Soviet Cybernetics and International Development," *The Social Impact of Cybernetics,* page 171.

98. Ibid.

99. Ibid.

100. *Computer Power and Human Reason: From Judgment to Calculation* (San Francisco: W. H. Freeman, 1976), page 156.

101. "The Organic Computer," *Discover* (May 1982), page 76.

VI THE NEW COSMIC MIRROR:
Finding a "Natural" Excuse for the Next World Epoch

1. "Artificial Synthesis of New Life Forms," *Bulletin of the Atomic Scientists,* 28 (December 1972), page 21.

2. Sayre, page 231.

3. Norbert Wiener, page 38.

4. Bjo Trimble, *The Star Trek Concordance* (New York: Ballantine Books, 1976), page 241.

5. Ibid.

6. Ibid.

7. Norbert Wiener, pages 130–31.

8. Ibid., page 131.

9. Ibid., page 140.

10. "Recombinant DNA: It's Not What We Need," March 7, 8, and 9, 1977.

11. "Whether to Make Perfect Humans," *The New York Times* (July 22, 1982), page A22.

12. Ibid.

13. Quoted in Albert Rosenfeld, *The Second Genesis* (New York: Vintage, 1975), page 212.

14. Quoted in Walter G. Peter III, "Ethical Perspectives in the Use of Genetic Knowledge," *Bio-Science,* 21 (1971), page 205.

15. Julian Huxley, *Evolution in Action* (New York: New American Library, 1953), page 31.

BIBLIOGRAPHY

BOOKS

Aitchison, Leslie. *A History of Metals.* New York: Interscience Publishers, 1960.

Altick, Richard D. *Victorian People and Ideas.* New York: W. W. Norton, 1973.

Ardrey, Robert. *The Territorial Imperative: A Personal Inquiry into the Animal Origins of Property and Nations.* New York: Atheneum, 1966.

Bachofen, J. J. *Myth, Religion and Mother Right.* Translated by Ralph Manheim. Princeton: Princeton University Press, 1973.

Barfield, Owen. *Saving the Appearances: A Study in Idolatry.* New York: Harcourt, Brace & World, 1965.

Barlow, Nora, ed. *Autobiography of Charles Darwin: 1809–1882.* New York: W. W. Norton, 1969.

Bateson, Gregory. *Mind and Nature: A Necessary Unity.* New York: E. P. Dutton, 1979.

———. *Steps to an Ecology of Mind.* Novato, Calif.: Chandler and Sharp, 1972/New York: Ballantine Books, 1972.

Beauvoir, Simone de. *The Second Sex.* Translated by H. M. Parshley. New York: Alfred A. Knopf, 1953.

Becker, Ernest. *Escape from Evil.* New York: The Free Press, 1975.

Beer, Gavin de. *Embryos and Ancestors.* London: Oxford University Press, 1954.

Bentov, Itzhak. *Stalking the Wild Pendulum: On the Mechanics of Consciousness.* New York: Dutton, 1977/Bantam Books, 1979.

Berman, Morris. *The Reenchantment of the World.* Ithaca: Cornell University Press, 1981.

Brown, Norman O. *Life Against Death: The Psychoanalytical Meaning of History.* Middletown, Conn.: Wesleyan University Press, 1959.

Bugliarello, George, and Koner, Dean B., eds. *The History and Philosophy of Technology.* Urbana, Ill.: University of Illinois Press, 1979.

Burckhardt, Titus. *Alchemy.* Translated by William Stoddart. London: Stuart & Watkins, 1967.

Burr, Harold Saxton. *Blueprint for Immortality: The Electric Patterns of Life.* London: Neville Spearman, 1972.

Butterfield, Herbert. *The Origins of Modern Science, 1300–1800.* New York: The Free Press, 1965.

Canetti, Elias. *Crowds and Power.* Translated by Carol Stewart. London: Gollancz, 1962.

Cantor, Norman F. *Medieval History: The Life and Death of a Civilization.* 2nd ed. New York: Macmillan, 1969.

Clark, Robert T., and Bales, James D. *Why Scientists Accept Evolution.* Grand Rapids, Mich.: Baker Book House, 1966.

Cobb, John, Jr., and Griffin, David, eds. *Mind in Nature: Essays on the Interface of Science and Philosophy.* Washington, D.C.: University Press of America, 1977.

Cohen, Mark Nathan. *The Food Crisis in Prehistory: Overpopulation and the Origins of Agriculture.* New Haven: Yale University Press, 1977.

Collingwood, R. G. *The Idea of Nature.* Oxford: Oxford University Press, 1945.

Conklin, Edwin G. *Man Real and Ideal: Observation and Reflections on Man's Nature, Development and Destiny.* New York: Scribner's, 1943.

Crick, Francis. *Life Itself: Its Origin and Nature.* New York: Simon & Schuster, 1981.

Daly, Herman E. *Steady-State Economics: The Economics of Biophysical Equilibrium and Moral Growth.* San Francisco: W. H. Freeman, 1977.

Darwin, Charles. *The Descent of Man and Selection in Relation to Sex.* New York: Appleton, 1896.

———. *The Life and Letters of Charles Darwin.* Edited by Francis Darwin. 2 vols. New York: Appleton, 1887.

———. *More Letters.* Edited by Francis Darwin and A. C. Steward. 2 vols. London: Dover Publications, 1903.

———. *The Origin of Species.* Introduction by Harrison Matthews. London: J. M. Dent, 1971.

————. *The Origin of Species.* New York: E. P. Dutton, 1956.

————. *The Origin of Species.* New York: Watts, 1929.

————. *The Origin of Species* and *The Descent of Man.* New York: Modern Library, 1936.

————. *The Variation of Animals and Plants Under Domestication.* 2 vols. New York: Appleton, 1896.

————. *The Voyage of the Beagle,* edited by Leonard Engel. Garden City, N.Y.: Doubleday, 1962.

Darwin, Charles, and Wallace, A. R. *Evolution by Natural Selection.* Cambridge, England: Cambridge University Press, 1958.

Dechert, Charles R., ed. *The Social Impact of Cybernetics.* New York: Simon & Schuster, 1967.

Deely, John N., and Nogar, Raymond J. *The Problem of Evolution: A Study of the Philosophical Repercussions of Evolutionary Science.* New York: Appleton-Century-Crofts, 1973.

Deloria, Vine, Jr. *The Metaphysics of Modern Existence.* New York: Harper & Row, 1979.

Dinnerstein, Dorothy. *The Mermaid and the Minotaur: Sexual Arrangements and Human Malaise.* New York: Harper & Row, 1976.

Dobzhansky, Theodosius; Ayala, Francisco J.; Stebbins, Ledyard, G.; and Valentine, James W. *Evolution.* San Francisco: W. H. Freeman, 1977.

Donoghue, Denis. *Thieves of Fire.* New York: Oxford University Press, 1974.

Durbin, Paul T. *A Guide to the Culture of Science, Technology, and Medicine.* New York: The Free Press, 1980.

Eiseley, Loren. *Darwin's Century.* New York: Doubleday, 1958.

————. *The Firmament of Time.* New York: Atheneum, 1960.

————. *The Immense Journey.* New York: Random House, 1957.

Eldredge, Niles, and Cracraft, Joel. *Phylogenetic Patterns and the Evolutionary Process: Method and Theory in Comparative Biology.* New York: Columbia University Press, 1961.

Eliade, Mircea. *The Myth of the Eternal Return.* Translated by Willard R. Trask. New York: Pantheon Books, 1954.

Ellul, Jacques. *The Technological Society.* Translated by John Wilkinson. New York: Alfred A. Knopf, 1964.

Falconer, Douglas Scott. *Introduction to Quantitative Genetics.* New York: Ronald Press, 1960.

Ferguson, Marilyn. *The Aquarian Conspiracy: Personal and Social Transformations in the 1980's.* Los Angeles: J. P. Tarcher, 1980.

Flaceliere, Robert. *Daily Life in Greece: At the Time of Pericles.* Translated by Peter Green. New York: Macmillan, 1965.

Fox, S. W., ed. *The Origins of Prebiological Systems and Their Molecular Matrices.* New York: Academic Press, 1965.

Fraser, J. T. *Of Time, Passion, and Knowledge: Reflections on the Strategy of Existence.* New York: George Braziller, 1975.

Freud, Sigmund. *Beyond the Pleasure Principle.* Translated by J. Strachey. New York: W. W. Norton, 1961.

Galton, Francis. *Inquiries into Human Faculty and Its Development.* New York: AMS Press, n.d.

Gillispie, Neal C. *Charles Darwin and the Problem of Creation.* Chicago: University of Chicago Press, 1979.

Gish, Duane T. *Evolution: The Fossils Say No!* San Diego: Creation-Life Publishers, 1978.

Glacken, Clarence J. *Traces on the Rhodian Shore.* Berkeley & Los Angeles: University of California Press, 1967.

Grassé, Pierre-P. *Evolution of Living Organisms: Evidence for a New Theory of Transformation.* New York: Academic Press, 1977.

Gray, Elizabeth Dodson. *Why the Green Nigger: Re-mything Genesis.* Wellesley, Mass.: Roundtable Press, 1979.

Greene, John C. *Science, Ideology, and World View: Essays in the History of Evolutionary Ideas.* Berkeley: University of California Press, 1981.

Grene, Marjorie. *The Understanding of Nature: Essays in the Philosophy of Biology.* Dordrecht, Holland: D. Reidel, 1974.

Gruber, Howard E., and Barrett, Paul H. *Darwin on Man: A Psychological Study of Scientific Creativity.* New York: E. P. Dutton, 1974.

Gunter, P. A. Y., ed. *Bergson and the Evolution of Physics.* Knoxville, Tenn.: University of Tennessee Press, 1977.

Gurvitch, Georges. *The Social Frameworks of Knowledge.* Translated by Margaret A. and Kenneth A. Thompson. Oxford: Basil Blackwell, 1971.

Hadd, John R. *Evolution: Reconciling the Controversy.* New Jersey: Kronos Press, 1979.

Haeckel, Ernst. *The Wonders of Life.* New York: Harper & Brothers, 1905.

Halacy, D. S., Jr. *The Genetic Revolution.* New York: Harper & Row, 1974.

Hall, Wilbur. *Partner of Nature.* New York: Appleton-Century-Crofts, 1939.

Hallpike, C. R. *The Foundations of Primitive Thought.* Oxford: Clarendon Press, 1979.

Hampden-Turner, Charles. *Maps of the Mind: Charts and Concepts of the Mind and Its Labyrinths.* New York: Macmillan, 1981.

Haraway, Donna Jeanne. *Crystals, Fabrics, and Fields: Metaphors of Organism in Twentieth-Century Developmental Biology.* New Haven: Yale University Press, 1976.

Hardin, Garrett. *Nature and Man's Fate*. New York: Rinehart, 1959/New American Library, 1961.

Harris, Marvin. *Cannibals and Kings: The Origins of Culture*. New York: Julian Press, 1963.

Heard, Gerald. *The Five Ages of Man: The Psychology of Human History*. New York: Julian Press, 1963.

Heer, Friedrich. *The Medieval World*. Translated by Janet Sondheimer. Cleveland: World Publishing, 1962/New York: New American Library, 1963.

Heidegger, Martin. *The Question Concerning Technology and Other Essays*. Translated by William Lovitt. New York: Harper & Row, 1977.

Himmelfarb, Gertrude. *Darwin and the Darwinian Revolution*. New York: W. W. Norton, 1959.

Hitching, Francis. *The Neck of the Giraffe: Where Darwin Went Wrong*. New Haven and New York: Ticknor & Fields, 1982.

Hobsbawm, E. J. *The Age of Capital: 1848–1875*. New York: Scribner's, 1975.

Hocart, A. M. *Social Origins*. London: Watts, 1954.

Hofstadter, Richard. *Social Darwinism in American Thought*. Rev. ed. Boston: Beacon Press, 1955.

Hoos, Ida R. *Systems Analysis in Public Policy*. Berkeley and Los Angeles: University of California Press, 1972.

Hopkins, Arthur John. *Alchemy: Child of Greek Philosophy*. New York: AMS Press, 1967.

Hoyt, Robert S. *Europe in the Middle Ages*. 2nd ed. New York: Harcourt, Brace & World, 1957.

Hues, J. Donald. *Ecology in Ancient Civilizations*. Albuquerque: University of New Mexico Press, 1975.

Huxley, Julian S. *Evolution in Action*. New York: Harper & Brothers, 1953/New American Library, 1957.

———. *Evolution, the Modern Synthesis*. London: Allen & Unwin, 1942.

Jantsch, Erich. *The Self-Organizing Universe: Scientific and Human Implications of the Emerging Paradigm of Evolution*. New York: Pergamon Press, 1980.

Jantsch, Erich, and Waddington, C. H., eds. *Evolution and Consciousness: Human System in Transition*. Reading, Mass.: Addison-Wesley, 1976.

Jaynes, Julian. *The Origin of Consciousness in the Breakdown of the Bicameral Mind*. Boston: Houghton Mifflin, 1976.

Johanson, Donald, and Edey, Maitland. *Lucy: The Beginnings of Humankind*. New York: Simon & Schuster, 1981.

Jung, C. G. *The Archetypes and the Collective Unconscious*. Translated by R. F. C. Hull. Princeton: Princeton University Press, 1969.

Jung, C. G., *et al. Man and His Symbols*. Garden City, N.Y.: Doubleday, 1964/Dell, 1968.

Kaplan, Bernard A. *Robots, Men and Minds*. New York: George Braziller, 1967.

Keosian, J. *The Origin of Life*. New York: Reinhold, 1968.

Kerkut, G. A. *Implications of Evolution*. New York: Pergamon Press, 1960.

Koestler, Arthur. *Janus: A Summing Up*. New York: Random House, 1978.

Kramer, Samuel Noah. *The Sumerians: Their History, Culture, and Character*. Chicago: University of Chicago Press, 1963.

Kumar, Satish. *The Schumacher Lectures*. New York: Harper & Row, 1980.

Lamberg-Karlovsky, C. C., ed. *Hunters, Farmers, and Civilizations*. San Francisco: W. H. Freeman, 1979.

Levy, Bernard-Henri. *The Testament of God*. Translated by George Holoch. New York: Harper & Row, 1980.

Lewis, John. *Beyond Chance and Necessity: A Critical Inquiry into Professor Jacques Monod's Chance and Necessity*. Brompton Road, London: Garnstone Press, 1974.

Lewis, R. W. B., ed. *Herman Melville*. New York: Dell, 1962.

Lovejoy, Arthur O. *The Great Chain of Being*. Cambridge, Mass.: Harvard University Press, 1936.

Lovelock, J. E. *Gaia: A New Look at Life on Earth*. Oxford: Oxford University Press, 1979.

Løvtrup, Søren. *Epigenetics: A Treatise on Theoretical Biology*. London: John Wiley & Sons, n.d.

McAuliffe, Sharon and Kathleen. *Life for Sale*. New York: Coward, McCann & Geoghegan, 1981.

Macbeth, Norman. *Darwin Retried: An Appeal to Reason*. Boston: Gambit, 1971.

McCulloch, John. *The Principles of Political Economy*. 2nd ed. London: Longman, Rees, Orme, Brown and Green, 1830.

McLuhan, Herbert Marshall. *Understanding Media: The Extensions of Man*. New York: McGraw-Hill, 1964.

McNeil, William. *The Rise of the West: A History of the Human Community*. Chicago: University of Chicago Press, 1963.

Margenau, Henry, and Sellon, Emily B., eds. *Nature, Man, and Society: Main Currents in Modern Thought*. York Beach, Me: Nicolas-Hays, 1976.

Marshall, A. J., ed. *Biology and Comparative Physiology of Birds*. 2 vols. New York: Academic Press, 1960.

Marx, Karl, and Engels, Friedrich. *Correspondence*. Translated and edited by Dona Torr. London: International Publishers, 1934.

————. *Selected Works.* 2 vols. London: Lawrence & Wishart Ltd., 1943.

————. *Selected Correspondence.* 2nd ed. Translated by I. Lasker. Moscow: Progress, 1965.

Maslow, Abraham H. *The Psychology of Science: A Reconnaissance.* New York: Harper & Row, 1966.

Mayr, Ernst. *Animal Species and Evolution.* Cambridge, Mass.: Harvard University Press, 1963.

————. *The Growth of Biological Thought.* Cambridge, Mass.: Harvard University Press, 1982.

Merleau-Ponty, Jacques, and Morando, Bruno. *The Rebirth of Cosmology.* Translated by Helen Weaver. New York: Alfred A. Knopf, 1976.

Mitcham, Carl, and Mackey, Robert, eds. *Philosophy and Technology: Readings in the Philosophical Problems of Technology.* New York: The Free Press, 1972.

Moorhead, Paul S., and Kaplan, Martin M., eds. *Mathematical Challenges to the Neo-Darwinian Interpretation of Evolution.* Philadelphia: The Wistar Institute Press, 1967.

Morris, Desmond. *The Naked Ape.* New York: McGraw-Hill, 1968/Dell, 1969.

Mulkay, Michael. *Science and the Sociology of Knowledge.* London: Allen & Unwin, 1979.

Mumford, Lewis. *Technics and Human Development: The Myth of the Machine.* Vol. I. New York: Harcourt, Brace & World, 1966.

Neuman, Erich. *The Origins and History of Consciousness.* New York: Pantheon Books, 1954.

Oakley, Francis. *The Medieval Experience: Foundations of Western Cultural Singularity.* New York: Scribner's, 1974.

Office of Technology Assessment. *Impacts of Applied Genetics: Micro-Organisms, Plants, and Animals.* Washington, D.C.: U.S. Government Printing Office, 1981.

Ommaney, F. D. *The Fishes.* New York: Time-Life, 1964.

Ong, Walter J. *Interfaces of the Word: Studies in the Evolution of Consciousness and Culture.* Ithaca: Cornell University Press, 1977.

Oparin, A. I. *Life: Its Nature, Origin, and Development.* Edinburgh: Oliver & Boyd, 1961.

Ophuls, William. *Ecology and the Politics of Scarcity.* San Francisco: W. H. Freeman, 1977.

Ord-Hume, Arthur W. J. G. *Perpetual Motion: The History of an Obsession.* New York: St. Martin's Press, 1977.

Ornstein, Robert E. *The Psychology of Consciousness.* New York: The Viking Press, 1973/Penguin Books, 1975.

Parker, Gary E. *Creation: The Facts of Life.* San Diego: Creation Life, 1980.

Piaget, Jean. *The Child's Conception of Time.* New York: Basic Books, 1969.

Popper, Karl R. *The Logic of Scientific Discovery.* New York: Harper & Row, 1965.

Prigogine, Ilya, and Stengers, I. *The New Alliance: Science in a Changing Context.* (In press.)

Radin, Paul. *The World of Primitive Man.* New York: Henry Schuman, 1953.

Rank, Otto. *Beyond Psychology.* New York: Dover Publications, 1941.

Rappaport, Roy A. *Ecology, Meaning, and Religion.* Richmond, Calif.: North Atlantic Books, 1979.

Roheim, Geza. *Psychoanalysis and Anthropology: Culture, Personality and the Unconscious.* New York: International Universities Press, 1968.

Rohlfing, D. L., and Oparin, A. I., eds. *Molecular Evolution: Prebiological and Biological.* New York: Plenum Press, 1972.

Romer, Alfred S. *Vertebrate Paleontology.* 3rd ed. Chicago: University of Chicago Press, 1966.

Rosnay, Joel de. *The Macroscope: A New World Scientific System.* Translated by Robert Edwards. New York: Harper & Row, 1979.

Ruse, Michael. *The Darwinian Revolution: Science Red Tooth and Claw.* Chicago: The University of Chicago Press, 1979.

Russell, Bertrand. *A History of Western Philosophy: And Its Connections with Political and Social Circumstances from the Earliest Times to the Present Day.* New York: Simon & Schuster, 1945.

———. *Religion and Science.* London: Oxford University Press, 1935.

Sachar, Abram Leon. *A History of the Jews.* 5th ed. New York: Alfred A. Knopf, 1964.

Sayre, Kenneth M. *Cybernetics and the Philosophy of Mind.* Highlands, N.J.: Humanities Press, 1976.

Simpson, George Gaylord. *The Meaning of Evolution.* New Haven: Yale University Press, 1967.

Smith, Adam. *An Inquiry into the Nature and Causes of the Wealth of Nations.* Oxford: Clarendon Press, 1976.

Smith, Huston. *Beyond the Post-Modern Mind.* New York: Crossroads, 1982.

Smith, Maynard J. *The Theory of Evolution.* 3rd ed. Harmondsworth, England: Penguin, 1975.

Sorokin, Pitirim. *Social and Cultural Dynamics.* Boston: Porter Sargent Publisher, 1970.

Spencer, Herbert. *Social Statics.* London: J. Chapman, 1851.

Spengler, Oswald. *The Decline of the West.* New York: Alfred A. Knopf, 1939.

Stauffer, R. C., ed. *Charles Darwin's Natural Selection.* Cambridge, England: Cambridge University Press, 1975.

Stent, Gunther S. *Paradoxes of Progress.* San Francisco: W. H. Freeman, 1978.

Stern, Karl. *The Flight from Women.* New York: Farrar, Straus & Giroux, 1965.

Symons, Donald. *The Evolution of Human Sexuality.* Oxford: Oxford University Press, 1979.

Teich, Mikulas, and Young, Robert, eds. *Changing Perspectives in the History of Science: Essays in Honor of Joseph Needham.* Boston: D. Reidal, 1973.

Thomson, John Arthur, and Geddes, Patrick. *Life: Outlines of General Biology.* 2 vols. London: Williams & Norgate, 1931.

Thomson, William Irwin. *The Time Falling Bodies Take to Light: Mythology, Sexuality, and the Origins of Culture.* New York: St. Martin's Press, 1981.

Thorpe, W. H., and Zangwill, O. L., eds. *Current Problems in Animal Behavior.* Cambridge: Cambridge University Press, 1961.

Tiger, Lionel, and Fox, Robin. *The Imperial Animal.* New York: Holt, Rinehart & Winston, 1971.

Toffler, Alvin. *Future Shock.* New York: Random House, 1970/Bantam Books, 1971.

———. *The Third Wave.* New York: Morrow, 1980/Bantam, 1981.

Trimble, Bjo. *The Star Trek Concordance.* New York: Ballantine Books, 1976.

Waddington, C. H. *The Evolution of an Evolutionist.* Edinburgh: Edinburgh University Press, 1975.

———. *The Strategy of the Genes.* London: Allen & Unwin, 1957.

Wallace, A. R. *Natural Selection and Tropical Nature.* London: Macmillan, 1895.

Weber, A. F. *The Growth of Cities in the Nineteenth Century.* New York: Macmillan, 1899.

Weiss, Paul A. *The Science of Life: The Living System—A System for Living.* New York: Futura, 1973.

Weizenbaum, Joseph. *Computer Power and Human Reason: From Judgment to Calculation.* San Francisco: W. H. Freeman, 1976.

Weizsacker, Carl Friedrich von. *The Unity of Nature.* Translated by Francis J. Zucker. New York: Farrar, Straus & Giroux, 1980.

Wells, H. G. *The Outline of History: The Whole Story of Man.* Vol. I. Revised by Raymond Postgate. Garden City, N.Y.: Garden City Books, 1961.

Wertime, Theodore A., and Muhly, James D. *The Coming of the Age of Iron.* New Haven: Yale University Press, 1980.

West, Geoffrey. *Charles Darwin: A Portrait.* New Haven: Yale University Press, 1938.

Whitehead, Alfred North. *Adventures of Ideas.* New York: Macmillan, 1937.

———. *The Concept of Nature.* Cambridge, England: Cambridge University Press, 1930.

———. *Nature and Life.* New York: Greenwood Press, 1968.

———. *The Principles of Natural Knowledge.* 2nd ed. Cambridge, England: Cambridge University Press, 1925.

Wiener, Martin J. *English Culture and the Decline of the Industrial Spirit: 1880–1980.* Cambridge: Cambridge University Press, 1981.

Wiener, Norbert. *God and Golem, Inc.: A Comment on Certain Points Where Cybernetics Impinges on Religion.* Cambridge, Mass.: M.I.T. Press, 1964.

———. *The Human Use of Human Beings: Cybernetics and Society.* Boston: Houghton Mifflin, 1954/New York: Avon Books, 1967.

Wilder-Smith, A. E. *The Natural Sciences Know Nothing of Evolution.* San Diego: Master Books, 1981.

Wilson, Edward. *On Human Nature.* Cambridge, Mass.: Harvard University Press, 1978/New York: Bantam Books, 1979.

Wilson, Peter J. *Man, The Promising Primate: The Conditions of Human Evolution.* New Haven: Yale University Press, 1980.

Worster, Donald. *Nature's Economy: The Roots of Ecology.* San Francisco: Sierra Club Books, 1977.

Wysong, R. L. *The Creation-Evolution Controversy.* Midland, Mich.: Inquiry Press, 1976.

ARTICLES

Alyea, Rosemary. "Researchers Change Sex of Lamb Fetus." *The Battalion,* February 1980.

Andrews, Lori B. "Inside the Genius Farm." *Parents Magazine,* October 1980.

Aschoff, Jürgen. "Circadian Rhythms in Man." *Science* 148 (1965).

Axelrod, Daniel I. "Early Cambrian Marine Fauna." *Science* 128 (July 4, 1958).

Beckwith, Jonathan. "Gene Expression in Bacteria and Some Concerns About the Misuse of Science." *Bacteriological Review* 34 (September 1970).

———. "Recombinant DNA: Does the Fault Lie within Our Genes?" Paper presented to the National Academy of Sciences Forum on Recombinant DNA, Washington, D.C., March 7, 8, and 9, 1977.

Bethell, Tom. "Darwin's Mistake." *Harper's,* February 1976.

"Biotechnology Markets to Soar: Efficiencies, Cost-Savings Seen." *Chemical Marketing Reporter,* August 25, 1980.

Bock, Walter J. Book review of *Der gerechtfertigte Haeckel and Nomogenesis, or Evolution Determined by Law. Science* 164 (1969).

"Bone Bonanza: Early Bird and Mastodon." *Science News,* September 24, 1977.

Bonner, John Tyler, "Implications of Evolution." *American Scientist* 49 (1961).

Bradie, Michael, and Gromko, Mark. "The Status of the Principle of Natural Selection." *Nature and System* 3 (1981).

Braithwaite, R. B.; Charlesworth, Deborah; Goodwin, Brian; Webster, Gerry; and Westphal, Jonathan. "Discussion: Neo-Darwinism." *Theoria to Theory* 13 (1979).

Brown, Frank A., Jr. "The Rhythmic Nature of Animals and Plants." *American Scientist* 47 (1959).

Bush, Paul. "Seed Firms Bought by Outside Interests." *Packer Weekly,* December 13, 1980.

"BW and Recombinant DNA." *Science Weekly,* April 1980.

Bylinsky Gene. "The Cloning Era Is Almost Here." *Fortune,* June 19, 1978.

Campbell, Colin. "What Happens When We Get . . . the Manchild Pill?" *Psychology Today,* August 1976.

Capek, Milič. "Temporal Order and Spatial Order: Their Differences and Relations." *Mind in Nature: Essays on the Interface of Science and Philosophy,* edited by John Cobb and David Griffin. Washington, D.C.: University Press of America, 1977.

Caplan, Arthur L. "Darwinism and Deductivist Models of Theory Structure." *Studies in the History and Philosophy of Science* 10 (1979).

Castelli, Jim. "Church Leaders Seek Probe of Gene Research." *The Washington Star,* June 21, 1981.

Chedd, Graham. "Danielli the Prophet." *New Scientist,* January 21, 1971.

Cohen, Mark Nathan. "Population Growth and Parallel Trends in Sociocultural Evolution." Paper presented to the 79th Annual Meeting of the American Anthropological Association, December 3–7, 1980, Washington, D.C.

Cohn, Victor. "Biologists Report Transfer of Gene from Rabbit to Mouse to Offspring." *The Washington Post,* September 8, 1981.

"Computer Age Began with Marvel of 1951." *The Washington Star,* June 10, 1981.

Condorcet, Marquis de. "Outline of an Historical View of the Progress of the Human Mind." *Main Currents in Modern Political Thought.* New York: Holt, Rinehart & Winston, 1950.

Cooke, Robert. "Biological Advances Have Business Applications." *The Boston Globe,* June 17, 1980.

"Counting Up the Genes." *The New York Times,* April 24, 1977.

Craig, Robin. "The Theoretical Possibility of Reverse Translation of Proteins into Genes." *Journal of Theoretical Biology* 88 (1981).

Crittenden, Ann. "Plan to Widen Plant Patents Stirs Conflict." *The New York Times,* June 6, 1980.

Danielli, James F. "Artificial Synthesis of New Life Forms." *Bulletin of Atomic Scientists* 28 (December 1972).

Danson, Roy. "Evolution." *New Scientist* 49 (1971).

Davidson, Keay. "Genetics: Evolution or Revolution?" *Orlando* (Florida) *Sentinel-Star,* March 22, 1981.

Davy, John. "What If Darwin Were Wrong?" *The Washington Post,* August 30, 1981.

Dechert, Charles R. "The Development of Cybernetics." *The Social Impact of Cybernetics,* edited by Charles R. Dechert. New York: Simon & Schuster, 1966.

Deevy, Edward S., Jr. "The Reply: Letter from Birnam Wood." *Yale Review* 61 (1967).

Dewey, John. "Metaphysical Issues: The Influence of Darwinism." *The Problem of Evolution,* edited by John Deely and Raymond Nogar. New York: Appleton-Century-Crofts, 1973.

Dickson, David. "Patenting Living Organisms—How to Beat the Bug-Rustlers." *Nature,* January 10, 1980.

"DNA Recombinant Molecule Research." Supplemental Report II, Report prepared for the Subcommittee on Science, Research, and Technology of the Committee on Science and Technology, House of Representatives, December 1976.

Dobzhansky, Theodosius. "On Methods of Evolutionary Biology and Anthropology." *American Scientist* 45 (1957).

Doolittle, W. Ford. "Is Nature Really Motherly?" *The Co-Evolution Quarterly* 29 (Spring 1981).

Doyle, Jack. "Green Revolution II: $." *The New York Times,* April 23, 1981.

"Dupont: Seeking a Future in Biosciences." *Business Week,* November 24, 1980.

Eckholm, Erik. "Disappearing Species: The Social Challenge." *Worldwatch Paper 22.* Washington, D.C.: Worldwatch Institute, June 1978.

Eden, Murray. "Inadequacies of Neo-Darwinian Evolution as a Scientific Theory." *Mathematical Challenges to the Neo-Darwinian Interpretation of Evolution,* edited by P. Moorhead and M. Kaplan. Philadelphia: Wistar Institute Press, 1967.

Ellegard, Alvara. "The Darwinian Theory and 19th Century Philosophies of Science." *Journal of the History of Ideas* 18 (1957).

Epps, Garrett. "Viroids Among Us." *Science 81,* September 1981.

Feder, Barnaby. "Automating Gene Splicing." *The New York Times,* January 15, 1981.

"Fetus for Sale." *Newsweek,* June 1, 1970.

Fiske, Edward B. "Computers Alter Life of Pupils and Teachers." *The New York Times,* April 4, 1982.

Ford, John J. "Soviet Cybernetics and International Development." *The Social Impact of Cybernetics,* edited by Charles R. Dechert. New York: Simon & Schuster, 1966.

Fox, Jeffrey, L. "Genetic Engineering Industry Has Growing Pains." *Chemical and Engineering News,* April 6, 1981.

Gale, Barry G. "Darwin and the Concept of a Struggle for Existence: A Study in the Extrascientific Origins of Scientific Ideas." *Isis* 63 (1972).

"Gene-Machine: Automated DNA Synthesis." *Science Weekly,* January 31, 1981.

"Gene Machines Will Add Players to the DNA Game." *Chemical Week,* February 4, 1981.

"Geneticist Says Recombinant DNA to Improve Cattle." *Feedstuffs Weekly,* June 2, 1980.

"Genetics Technology May Greatly Change Drug, Food Industries in the Next 20 Years." *The Wall Street Journal,* April 22, 1981.

Ghiselin, Michael T. "The Metaphysics of Phylogamy," review of *Phylogenetic Patterns and the Evolutionary Process* by Niles Eldredge and Joel Cracaft. *Paleobiology* 7 (1981).

Goldschmidt, R. B. "Evolution, As Viewed by One Geneticist." *American Scientist* 40 (1952).

Goodell, Rae. "The Gene Craze." *Columbia Journalism Review,* November/December 1980.

Goodwin, B. C. "The Cell as a Resonating System." *Towards a Theoretical Biology,* edited by C. H. Waddington. 2 vols. Chicago: Aldine, 1969.

———. "On Morphogenetic Fields." *Theoria to Theory* 13 (1979).

Goodwin, B. C., and Pateromichelakis, S. "The Role of Electrical Fields, Ions, and the Cortex in the Morphogenesis of Acetabularia." *Planta* 145 (1979).

Goodwin, B. C., and Trainor, L. E. H. "A Field Description of the Cleavage Process in Embryogenesis." *Journal of Theoretical Biology* 86 (1980).

Gorczynski, R. M., and Steele, E. J. "Simultaneous Yet Independent Inheritance of Somatically Acquired Tolerance to Two Distinct H-2 Antigenetic Haplotype Determinants in Mice." *Nature,* February 19, 1981.

Gould, Stephen Jay. "The Evolutionary Biology of Constraint." *Daedalus* 29 (1980).

———. "Is a New and General Theory of Evolution Emerging?" *Paleobiology* 6 (Winter 1980).

———. "The Return of Hopeful Monsters." *Natural History,* June/July 1977.

Gould, Stephen J., and Eldredge, Niles. "Punctuated Equilibria: The Tempo and Mode of Evolution Reconsidered." *Paleobiology* 3 (Spring 1977).

Gregory, R. L. "The Brain as an Engineering Problem." *Current Problems in Animal Behavior,* edited by W. H. Thorpe and O. L. Zangwill. Cambridge: Cambridge University Press, 1961.

Gurwitsch, Aron. "Comment on the Paper by H. Marcuse." *Boston Studies in the Philosophy of Science.* 2 vols. New York: Humanities Press, 1965.

Hahlweg, Kai. "Progress through Evolution? An Inquiry into the Thought of C. H. Waddington." *Acta Biotheoretica* 30 (1981).

Hartley, Brian. "The Biology Business." *Nature,* January 10, 1980.

Hartshorne, Charles. "Physics and Psychics: The Place of Mind in Nature." *Mind in Nature: Essays on the Interface of Science and Philosophy,* edited by John Cobb and David Griffin. Washington, D.C.: University Press of America, 1977.

Hawkes, Jacquetta and Christopher. "Land and People." *The Character of England,* edited by Ernest Baker. Oxford: Clarendon Press, 1947.

Hilts, Philip. "The Clock Within." *Science,* December 1980.

Ho, M. H., and Saunders, P. T. "Beyond Neo-Darwinism—An Epigenetic Approach to Evolution." *Journal of Theoretical Biology* 78 (1979).

"How Human Life Begins." *Newsweek,* January 11, 1982.

Huxley, Julian. "At Random—A Television Preview." *Evolution After Darwin,* edited by Sol Tax. 2 vols. Chicago: University of Chicago Press, 1960.

"Impacts of Applied Genetics: Microorganisms, Plants, and Animals." *Office of Technology Assessment,* U.S. Government Printing Office, Washington, D.C.

"Is Genetic Engineering Being Applied to Biological Warfare?" *Federation of American Scientists' Public Interest Report,* June 1980.

Jackson, David. "Is Genetic Engineering Safe?" *Baltimore Sun,* November 8, 1980.

King, Jonathon. "New Diseases in New Niches." *Nature,* November 2, 1978.

———. "Recombinant DNA and Auto Immune Disease." *The Journal of Infectious Diseases* 137 (1978).

Kitts, David B. "Paleontology and Evolutionary Theory." *Evolution* 28 (1974).

Knox, Richard A. "Life-forms Patents: Effects May Be Far Reaching." *Boston Globe,* June 29, 1980.

Krimsky, Sheldon. "Is Genetic Engineering Safe?" *Baltimore Sun,* November 8, 1980.

Krohn, Ingrid M. "Roles of Ideas in Advancing Paleontology." *Paleontology* 5 (1979).

Kung, Joan. "Some Aspects of Form in Aristotle's Biology." *Nature and System* 2 (1980).

Lancaster, Hal. "Gene Splicers Ponder Mass Production: Will Fermenting Pose Major Problems?" *The Wall Street Journal*, August 28, 1981.

Lessing, Lawrence. "Into the Core of Life Itself." *Fortune*, March 1966.

Lewin, Roger. "Biggest Challenge Since the Double Helix." *Science*, April 3, 1981.

———. "Evolutionary Theory Under Fire." *Science*, November 21, 1981.

———. "How Conversational Are Genes?" *Science*, April 17, 1981.

Løvtrup, Søren. "The Evolutionary Species: Fact or Fiction." *Systematic Zoology* 28 (1979).

McAuliffe, Kathleen. "Biochip Revolution." *Omni*, March 3, 1982.

Manier, Edward. "History, Philosophy and Sociology of Biology: A Family Romance." *Studies in the History and Philosophy of Science* 2 (1980).

Mayr, Ernst. "Evolution." *Scientific American*, September 1978.

Meltzer, Yale L. "Genetic Engineering: Building New Profits." *Chemical Marketing Reporter*, April 6, 1981.

Menosky, Joseph A. "Cheap, Fast Designer Genes: Machines That Allow Almost Anybody to Create New Life." *The Washington Post*, September 6, 1981.

———. "The Gene Machine." *The Washington Post*, September 6, 1981.

Merril, C. R., et al. "Bacterial versus Gene Expression in Human Cells." *Nature* 233 (1971).

Miller, Stanley. "Production of Some Organic Compounds Under Possible Primitive Conditions." *Journal of the American Chemical Society* 77 (1955).

Milne, David H. "How to Debate with Creationists—and 'Win.' " *The American Biology Teacher* 43 (1981).

Mitcham, Carl. "Philosophy of Technology." *Science, Technology, Medicine*, edited by Paul T. Durbin. New York: The Free Press, 1980.

Mitchell, John. "The Ideal World View." *The Schumacher Lectures*, edited by Satish Kumar. New York: Harper & Row, 1981.

Monro, Robert E. "Interpreting Molecular Biology." *Beyond Chance and Necessity*, edited by John Lewis. London: Garnstone Press, 1974.

Moore, John N. "Paleontologic Evidence and Organic Evolution." *Origins and Change*, edited by David L. Willis. Elgin, Ill.: A.S.A., 1978.

Mora, P. "The Folly of Probability." *The Origins of Prebiological Systems and Their Molecular Matrices*, edited by S. W. Fox. New York: Academic Press, 1965.

Morowitz, Harold J. "Reducing Life to Physics." *The New York Times*, June 23, 1980.

Newark, Peter. "The Origins of Biotechnology." *Nature,* January 10, 1980.

"New Genetic Technologies Willl Have Major Impact on Pharmaceutical, Chemical, Food and Processing Industries." Office of Technology Assessment News Release, April 22, 1981.

Nuttal, Tony. "Neo-Darwinism." *Theoria to Theory* 13 (1979).

"The Organic Computer." *Discover,* May 3, 1982.

Olson, Everett Claire. "The Evolution of Life." *Evolution After Darwin,* edited by Sol Tax. 2 vols. Chicago: University of Chicago Press, 1960.

O'Toole, Thomas. "In the Lab: Bugs to Grow Wheat, Eat Metal." *The Washington Post,* June 18, 1980.

Palmer, John D. "The Many Clocks of Man." *Cycles,* February 1971.

Perlman, David. "Biochemist's Attack on Gene Splicers." *San Francisco Chronicle,* November 15, 1978.

Prial, Frank J. "Pope, Ending His Visit to France, Warns of Danger of Nuclear War." *The New York Times,* June 3, 1980.

Randolph, Eleanor. "Bean-Squawk—Seed Patents: Fears Sprout at Grass Roots." *Los Angeles Times,* June 2, 1980.

Raup, David. "Conflicts Between Darwin and Paleontology." *Field Museum of Natural History Bulletin,* January 1979.

Rawls, Rebecca. "Nitrogen Fixation Research Advances." *Chemical and Engineering News,* December 8, 1980.

Riedel, Rupert. "A Systematic-Analytical Approach to Macro-Evolutionary Phenomena." *The Quarterly Review of Biology* 52 (1977).

Ritchie-Calder, Lord. "Retailoring the Tailor." *1976 Encyclopaedia Britannica, Book of the Year,* Special Supplement.

Rosenberg, Barbara, and Simon, Lee. "Recombinant DNA: Have Recent Experiments Assessed All the Risks?" *Nature,* December 27, 1979.

Rosenblueth, A.; Wiener, N.; and Bigelow, J. "Behavior, Purpose and Teleology." *Philosophy and Science* 10 (1943).

Rosenburg, Alexander. "Species Notions and the Theoretical Hierarchy of Biology." *Nature and System* 2 (1980).

Ruse, Michael. "Darwin's Debt to Philosophy: An Examination of the Influence of the Philosophical Ideas of John F. W. Hershel and William Whewell on the Development of Charles Darwin's Theory of Evolution." *Studies in the History and Philosophy of Science* 6 (1975).

———. "Natural Selection in the 'Origin of Species.' " *Studies in the History and Philosophy of Science* 1 (1971).

Sandow, Alexander. "Social Factors in the Origin of Darwinism." *The Quarterly Review of Biology* 13 (1938).

Saunders, P. T., and Ho, M. W. "On the Increase in Complexity of Evolution." *Journal of Theoretical Biology* 63 (1976).

Schmeck, Harold M., Jr. "Justices' Ruling Recognizes Gains in the Manipulation of Life Forms." *The New York Times,* June 17, 1980.

Schutzenberger, Marcel P. "Algorithms and the Neo-Darwinian Theory of Evolution." *Mathematical Challenges to the Neo-Darwinian Interpretation of Evolution,* edited by P. Moorhead and M. Kaplan. Philadelphia: Wistar Institute Press, 1967.

Schweber, Silvan S. "Darwin and the Political Economists." *Journal of the History of Biology* 13 (1980).

————. "The Origin of the 'Origin' Revisited." *Journal of the History of Biology* 10 (1977).

Scott, Cynthia. "Genetic Engineering: Menace or Mandate from God." *Moody,* March 1981.

Severo, Richard. "Genetic Tests by Industry Raise Questions on Rights of Workers." *The New York Times,* February 3, 1980.

Sheldrake, Rupert. "A New Science of Life." *New Scientist,* June 18, 1981.

Shrader, Douglas. "The Evolutionary Development of Science." *The Review of Metaphysics* 43 (1981).

Shurkin, Joel N. "Yet Another Step in the Complex Probe of the Genetic Code." *Philadelphia Inquirer,* May 1, 1977.

Signer, Ethan. "Recombinant DNA: It's Not What We Need." Written remarks presented at the NAS Forum, March 7, 8, and 9, 1977.

Simpson, George Gaylord, ed. "Notes on the Nature of Science by a Biologist." *Notes on the Nature of Science,* edited by George Gaylord Simpson et al. New York: Harcourt, Brace & World, 1962.

Smith, T. F., and Waterman, M. S. "Overlapping Genes and Information Theory." *Journal of Theoretical Biology* 91 (1981).

Sullivan, Walter. "Creation Debate Is Not Limited to Arkansas Trial." *The New York Times,* December 27, 1981.

Szent-Gyorgyi, A. "The Evolutionary Paradox and Biological Stability." *Molecular Evolution: Prebiological and Biological,* edited by D. L. Rohlfing and A. I. Oparin. New York: Plenum Press, 1972.

Tamarkin, Bob. "The Growth Industry." *Forbes,* March 2, 1981.

"Test Frozen Embryos for Invitro Babies." *Sexual Medicine Today,* January 1980.

Theobald, Robert. "Cybernetics and the Problems of Social Reorganization." *The Social Impact of Cybernetics,* edited by Charles R. Dechert. New York: Simon & Schuster, 1966.

Thompson, Keith Stewart. "Something in Common," review of *Phylogenetic Patterns and the Evolutionary Process* by Niles Eldredge and Joel Cracraft. *Paleobiology* 7 (1981).

Thorpe, W. H. "The Frontiers of Biology." *Mind in Nature,* edited by John Cobb and David Griffin. Washington, D.C.: University Press of America, 1977.

"Too Soon for the Rehabilitation of Lamarck." *Nature,* February 19, 1981.

Tudge, Colin. "Lamarck Lives—In the Immune System." *New Scientist,* February 19, 1981.

Tyson, Kim. "Genetic Maps at Hand, Scientist Says." *American-Statesman,* December 3, 1980.

Vorzimmer, Peter J. "Darwin's Question About the Breeding of Animals (1839)." *Journal of the History of Biology* 2 (1969).

———. "An Early Darwin Manuscript: The 'Outline and Draft of 1839.' " *Journal of the History of Biology* 8 (1975).

Waddington, C. H. "A Catastrophe Theory of Evolution." *Annals of the New York Academy of Science* 231 (1974).

———. "Whitehead and Modern Science." *Mind in Nature,* edited by John Cobb and David Griffin. Washington, D.C.: University Press of America, 1977.

Wade, Nicholas. "Getting Rich in the Ivory Tower: Hot Genes." *The New Republic,* November 1980.

Walgate, Robert. "How Safe Will Biobusiness Be?" *Nature,* January 10, 1980.

White, Errol. "Proceedings." *Linnaean Society of London* 177 (1966).

Willis, David L. "Creation and/or Evolution." In *Origins and Change,* edited by David L. Willis. Elgin, Ill.: A.S.A., 1978.

Winfree, Arthur. "Chemical Clocks: A Clue to Biological Rhythms." *New Scientist,* October 5, 1978.

"A Working Synthetic Gene." *Medical World News,* September 20, 1976.

Yanchinski, Stephanie. "DNA: Ignorant, Selfish and Junk." *New Scientist,* July 16, 1981.

Yockey, Hubert P. "A Calculation of the Probability of Spontaneous Biogenesis by Information Theory." *Journal of Theoretical Biology* 67 (1977).

———. "Rebuttal of 'Overlapping Genes and Information Theory.' " *Journal of Theoretical Biology* 91 (1981).

———. "Self-Organization Origin of Life Scenarios and Information Theory." *Journal of Theoretical Biology* 91 (1981).

Young, Robert M. "Darwinism and the Division of Labour." *Listener,* August 17, 1972.

———. "Malthus and the Evolutionists: The Common Context of Biological and Social Theory." *The Past and Present Conference* 43 (1969).

———. "Man's Place in Nature." *Changing Perspectives in the History of Science,* edited by Mikulus Teich and Robert Young. Boston: D. Reidal, 1978.

INDEX

Jeremy Rifkin is the director of the Foundation on Economic Trends, based in Washington, D.C., which has sponsored his *Who Should Play God?*, *The North Will Rise Again*, *The Emerging Order*, and the best-selling *Entropy*.